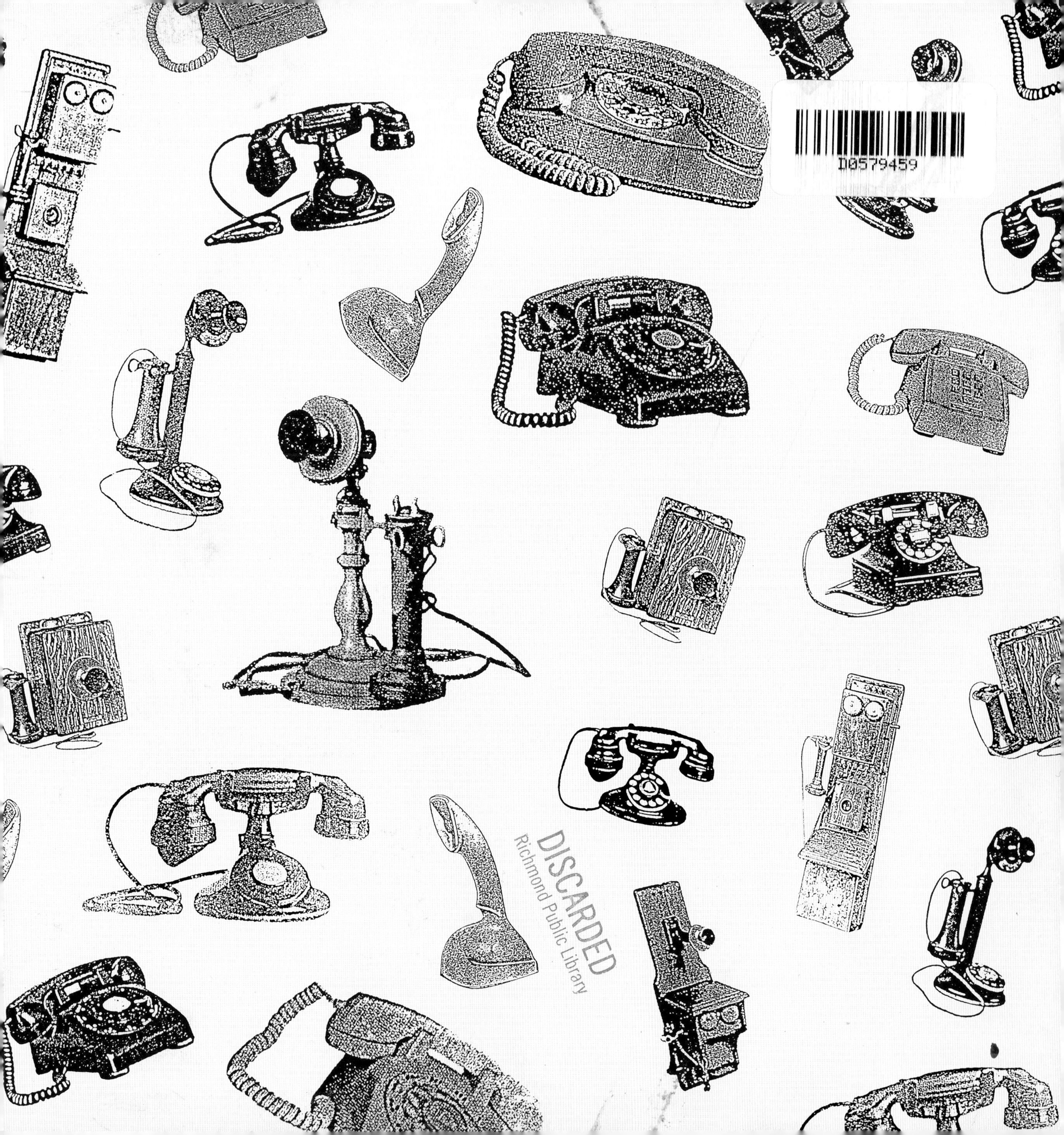
D0579459
DISCARDED
Richmond Public Library

Once Upon a Telephone

An Illustrated Social History

Once Upon a Telephone

An Illustrated Social History

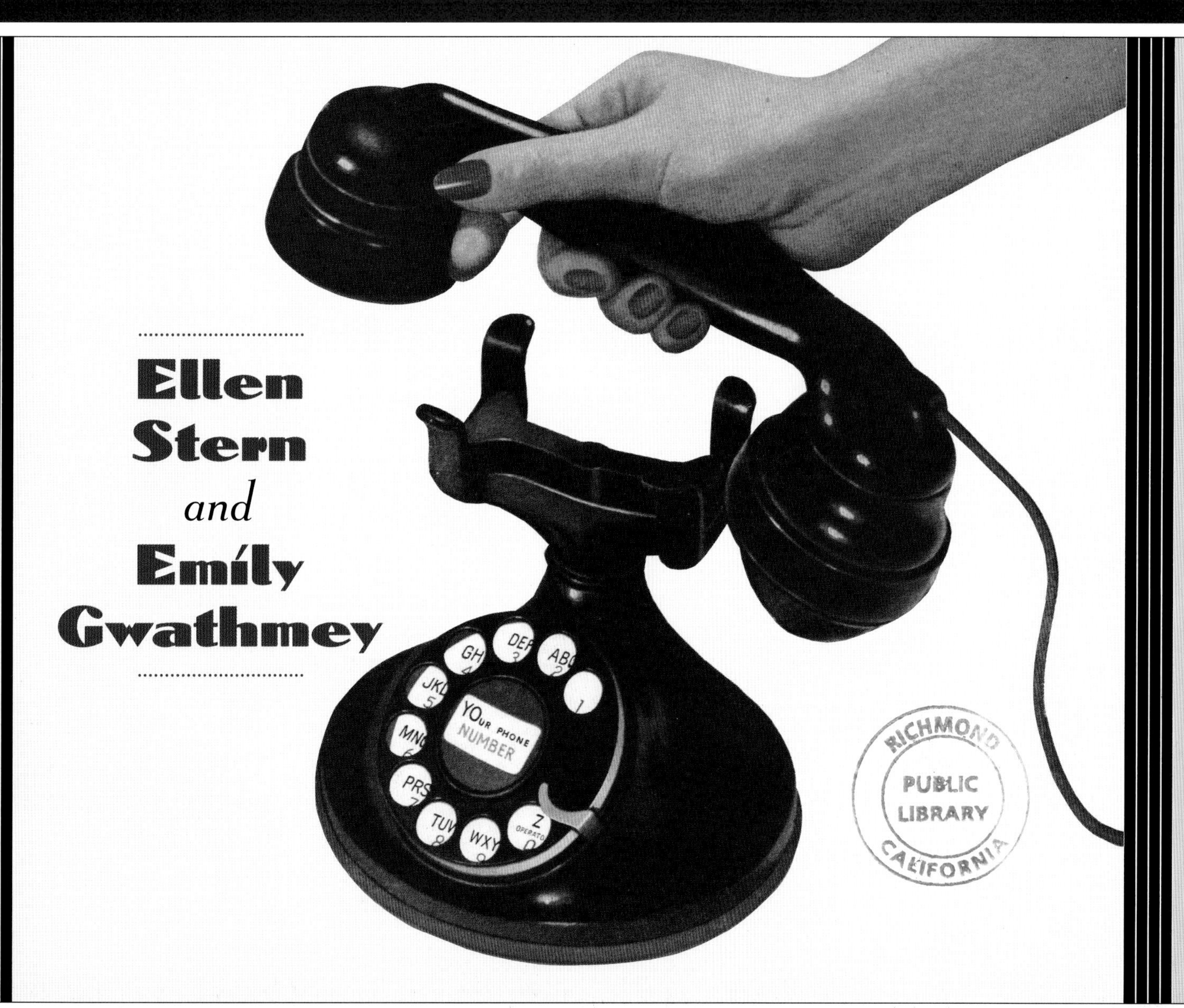

Ellen Stern

and

Emily Gwathmey

HARCOURT BRACE & COMPANY

New York San Diego London

HARCOURT
BRACE

Library of Congress Cataloging-in-Publication Data
Stern, Ellen Stock.
Once upon a telephone: an illustrated social history/Ellen Stern and Emily Gwathmey.
p. cm.
ISBN 0-15-100086-7
1. Telephone—Social aspects—United States—History.
2. Telephone—United States—Anecdotes. I. Gwathmey, Emily Margolin. II. Title
HE8817.S86 1994
302.23'5'0973—dc20 93-50854

Designed by G. B. D. Smith
Printed in Singapore
First edition
A B C D E

Permissions and acknowledgments appear on pages 134–135, which constitute a continuation of the copyright page.

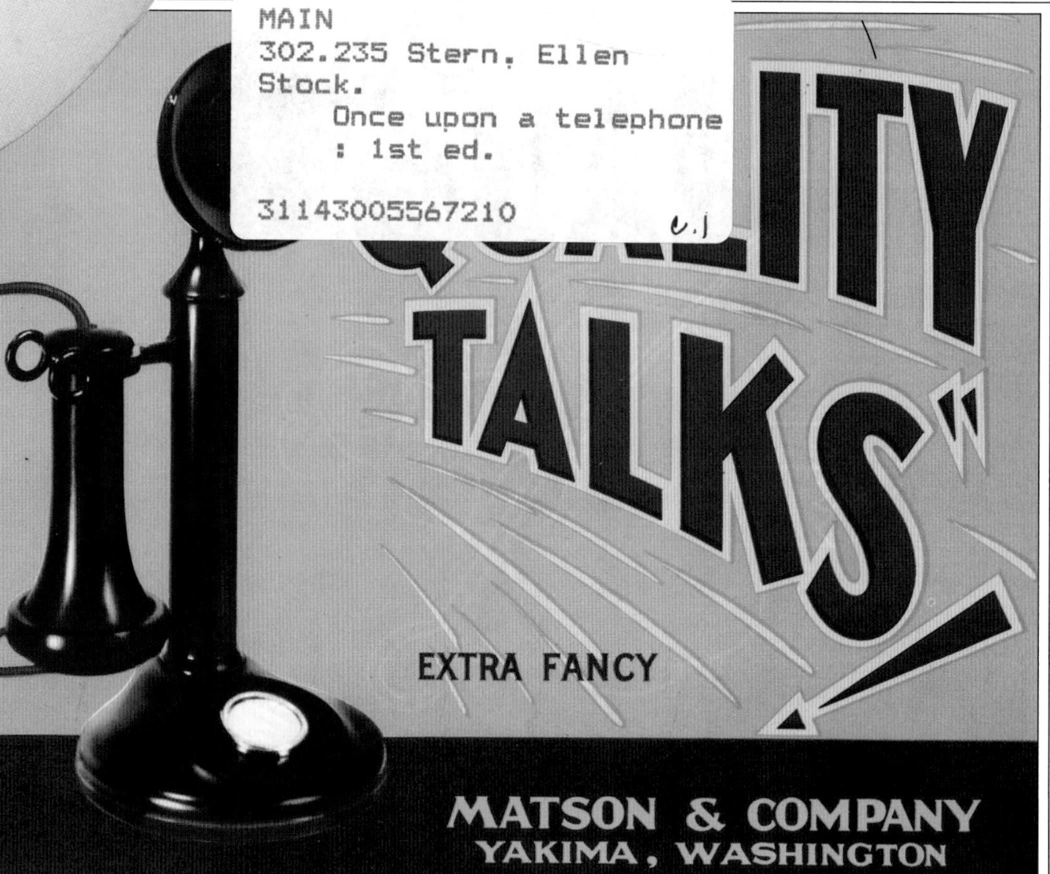

Acknowledgments

We are deeply grateful to Robert Breish, Sandy Marrone, Dan Goldin, and Neil Breslin for sharing with us their telephoniana, expertise, sustenance, and good cheer. And to Martin Eisen, for his invaluable library. Thanks, too, to Jim Kilpatric and Marcy Goldstein at AT&T, Neil Connors and Peter Turck at NYNEX, Jill Denman at the Yellow Pages Publishers Association, Margaret A. Allyn at the Plaza, Diane Wall at Hallmark, William G. Seekamp and Susan Waskevich at Southern New England Telephone, Russell A. Flinchum at the Cooper-Hewitt Museum, Barbara Strauch, Faith Coleman, Thom Todd, Ted Lee, our designer G. B. D. Smith, our agent Lisa Bankoff and our editor Claire Wachtel. And to the New York Society Library, our gratitude for its splendid books, staff, elegance, existence.

TELEPHONE BLUES
(THOSE DOGGONE TELEPHONE BLUES)
WORDS AND MUSIC BY
CHRISTIAN MARCUS
LOVELIGHT MUSIC
CO
1547 Broadway, New York
MADE IN U.S.A.
Politzer

Directory

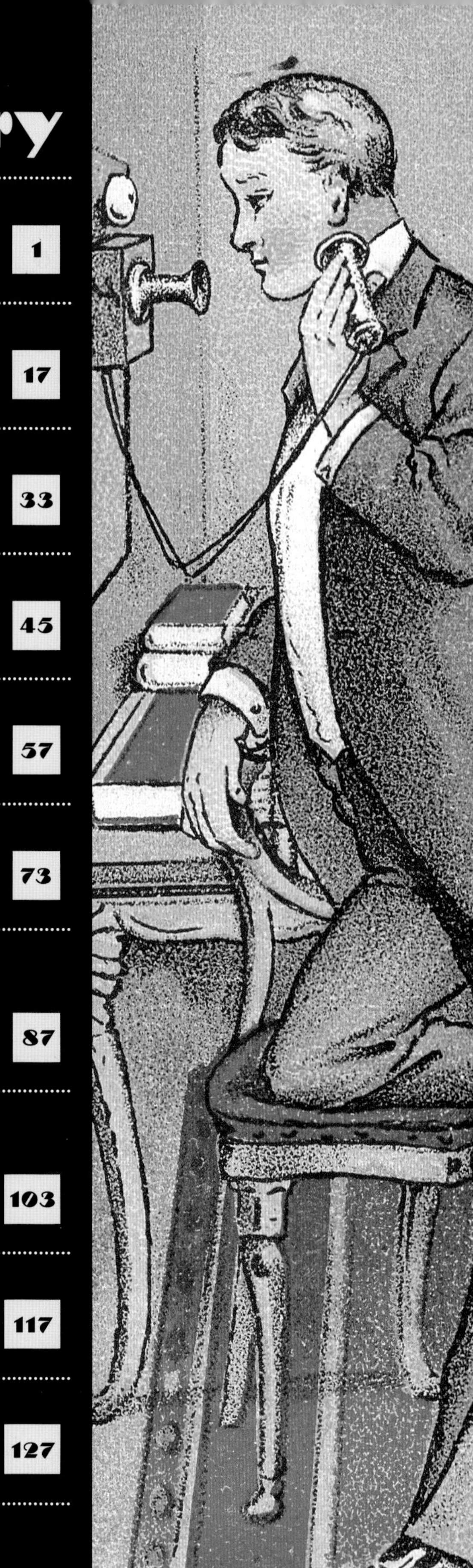

Chapter 1

"My God, It Talks!"

(Above) *"Mr. Watson. Come here! I want you!" Bell called over the Liquid Telephone of 1876.* (Above right) *The Centennial model.* (Opposite) *Bell himself on a later version.*

Alexander Graham Bell was born on the morning of the electrical age into a family whose lives and passions revolved around sound, speech, and communication. His grandfather, Alexander Bell, was a golden-voiced Scotsman of the early nineteenth century. A cobbler, tavern keeper, and comedian who also gave staged readings, he gained prominence as a professor of elocution, turning out two books on the subject. In London, he opened an elocution school and developed a lucrative practice in the treatment of stammering, lisping, and other speech disorders. Legend has it that his expertise in smoothing out a Cockney accent so impressed George Bernard Shaw that he became the model for Professor Henry Higgins in *Pygmalion.*

His father, Alexander Melville Bell, was a linguist who taught speech to the deaf, wrote textbooks, and gained international fame by inventing a phonetic alphabet called Visible Speech. This coded system reduced all verbal sounds to thirty-four written sym-

Alexander Graham Bell with his wife, Mabel, and daughters Elsie May (left) and Marian ("Daisy") in December, 1885.

bols; the symbols indicated placement of the tongue, throat, and lips during speech. Its original purpose was to aid in the pronunciation of foreign languages. Surprisingly, it turned out to be most effective in teaching deaf people to speak intelligibly. Conflicting legend has it that it was Melville Bell's ingenious Visible Speech that inspired Shaw's Professor Higgins.

Alexander Graham Bell's mother, Eliza Symonds Bell, was deaf. A painter of miniatures, she was also a talented pianist who was able to "hear" the music she played by placing the mouthpiece of her ear tube on the sounding board. With a tuning fork in each hand, her young son would stand for hours toying with the piano's vibrating strings and the way they affected the tongs. The mysterious nature of sound and how it travels along wires and into the air would reverberate in his mind and intrigue him all his life.

Central casting couldn't have done better.

ALEXANDER GRAHAM BELL CAME INTO THIS REMARKABLE household on March 3, 1847. Introspective and often withdrawn, Bell inherited his love of music from his mother (he, too, played the piano beautifully) and a resonant speaking voice from his father. He was tutored at home, then attended Edinburgh's Royal High School. Reading and writing poetry interested him, but school did not. He much preferred collecting plants, dissecting animals, and tinkering with tuning forks. Ever curious and piqued by challenge, he concocted for a local farmer a device to remove husks from wheat by combining a nail brush and paddle into a rotary brushing wheel. He was fourteen.

At fifteen, Aleck's parents sent him to London to acquire polish. Under the influential wing of Grandfather Bell, he learned how to dress, how to behave in proper society, and how to speak before an audience when the two gave public performances reading Shakespeare together.

During a visit to the London workshop of Sir Charles Wheatstone, Aleck was fascinated by one of the scientist's inventions: a speaking machine. Back home, he and his older brother Melville tried one of their own with a facsimile skull and homemade jaws, teeth, larynx, lips, cheeks, tongue, and palate. With a bellows, the machine was able to cry "Ma-ma." Going even further with his domestic experiments, Aleck figured out a way to manipulate the mouth and vocal cords of his Skye terrier so that the dog's growls actually sounded like words.

Curious and ambitious, he set out early on the path of instruction, following in the family footsteps. At sixteen, he took a position teaching music and elocution at a boys' boarding school. After further study at the University of Edinburgh and the University of London, he and his two brothers traveled all over Scotland to demonstrate to incredulous audiences their father's Visible Speech.

Elsewhere in the world, the German physicist Hermann von Helmholtz had also been experimenting with sound. His thesis, *On the Sensations of Tone,* proved that vowel sounds could be produced not only by the human mouth but also by a combination of electrical tuning forks and resonators. Bell could not read German, but he tried anyway; confusing diagrams further befuddled him. And so, incorrectly, he understood Helmholtz to conclude not just that vowels could be reproduced electromagnetically, but that they could be transmitted from here to there over a wire.

"It was a very valuable blunder," Bell later admitted. "It gave me confidence. If I had been able to read German, I might never have begun my experiments in electricity." Nor landed on the path to telephonic glory.

The Bells moved to London. Then, in 1870, tragedy struck. Aleck's two brothers, Melville and Edward, died of tuberculosis within four months of each other. Fearing further loss, since Aleck, twenty-three, was already suffering from it himself, his parents packed him up and sailed away to safe haven in Canada.

They settled in a spacious farmhouse in Brant-

Bell's "dreaming place": the Canadian home where creativity flowered.

DARRYL F. ZANUCK'S
PRODUCTION OF
THE STORY OF
ALEXANDER
GRAHAM BELL
DIRECTED BY IRVING CUMMINGS
ASSOCIATE PRODUCER KENNETH MACGOWAN
DON
AMECHE
LORETTA
YOUNG
HENRY
FONDA

ford, Ontario, where Aleck recuperated. "This is my dreaming place!" he wrote of the spot he most loved, high on a cliff overlooking the Grand River. This was his retreat after restless bouts of activity and exploration. He cherished solitude—but not for too long. For fun, he learned the language of the nearby Mohawks, and learned it so well he was initiated into the tribe. All the while, he fiddled and experimented with sound and electricity.

Teaching deaf children to speak was a revolutionary idea in America, where sign language or institutionalization were the prevailing remedies. Disagreeing with these exclusionary methods, Aleck traveled to America as his father's emissary and in 1872 introduced Visible Speech at the Boston School for Deaf Mutes. Under the tutelage of this miracle worker, the deaf began to speak.

He lectured. He founded and edited a periodical called *Visible Speech Pioneer.* He opened his own school of vocal physiology and the mechanics of speech. And he became a professor at Boston University.

It was at this time that he met two prominent men who would become his lifelong supporters and champions. One was Thomas Sanders, a wealthy Salem leather-and-hides merchant, whose deaf son, George, would become Bell's student. The other was Gardiner Greene Hubbard, a prominent patent attorney and president of the Clarke School for the Deaf in Northampton. Hubbard's daughter, Mabel, had lost her hearing at the age of four during an epidemic of scarlet fever.

When not teaching, Bell turned his attention to Samuel F. B. Morse's telegraph, which had been considered a miracle wrought by God when introduced in 1844. But, despite a web of wires crisscrossing the country, telegraphy was limited. It required operators to translate coded messages sent over a wire, and only one message could be transmitted at a time. This caused untold backups and delays at telegraph offices across the country.

Competition to improve the situation was fierce. Bell and his many rivals—among them, Thomas A. Edison and the Western Union Telegraph Company—were striving to develop a multiple telegraph, which would be capable of sending several messages over a single wire simultaneously. Since he recognized the parallel between multiple messages and the multiple notes in a musical chord, Bell called his dream the "harmonic telegraph." Sanders and Hubbard bought the dream—which, if practicable, would make them all unbelievably rich.

Enter Mr. Watson.

The electrical machine shop of Charles Williams at 109 Court Street was a-hum with the workings of burglar alarms, annunciators, galvanometers, telegraph keys, sounders, relays, registers, and printing telegraph instruments. It was to this electric mecca that Bell came in 1874 with the blueprints for his harmonic telegraph. His idea was to replace his tuning forks with flexible organ reeds that would be attached to electromagnets, but he was not mechanically clever and needed help in putting the thing together. He found Thomas A. Watson, prize artisan.

As the industrious Watson remembered their historic meeting—which occurred one wintry day as he was toiling over torpedo parts in the Boston shop—"a tall, slender, quick-motioned man with pale face, black side whiskers, and drooping mustache, big nose and high sloping forehead crowned with bushy, jet black hair, came rushing out of the office and over to my

Twentieth Century-Fox's 1939 movie classic inspired a popular nickname for the telephone: "the Ameche."

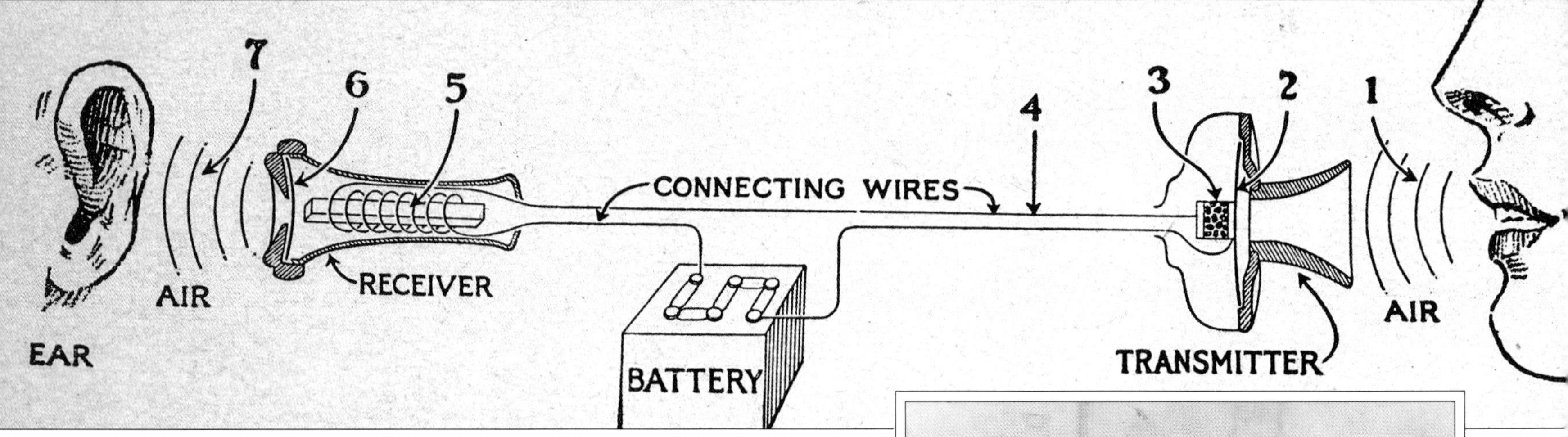

workbench. It was Alexander Graham Bell whom I saw then for the first time." Bell's harmonic telegraph, Watson recalled in an address delivered before the Third Annual Convention of the Telephone Pioneers of America at Chicago, in 1913, "was a simple affair by means of which, utilizing the law of sympathetic vibration, he expected to send six or eight Morse messages on a single wire at the same time, without interference."

All winter they tried, but they couldn't get it to work. "One evening when we were resting from our struggles with the apparatus," Watson went on, "Bell said to me: 'Watson, I want to tell you of another idea I have which I think will surprise you.' . . . but when he went on to say that he had an idea by which he believed it would be possible to talk by telegraph, my nervous system got such a shock that the tired feeling vanished. I have never forgotten his exact words; they have run in my mind ever since like a mathematical formula.

" 'If,' he said, 'I could make a current of electricity vary in intensity, precisely as the air varies in density during the production of a sound, I should be able to transmit speech telegraphically.' He then sketched for me an instrument that he thought would do this, and we

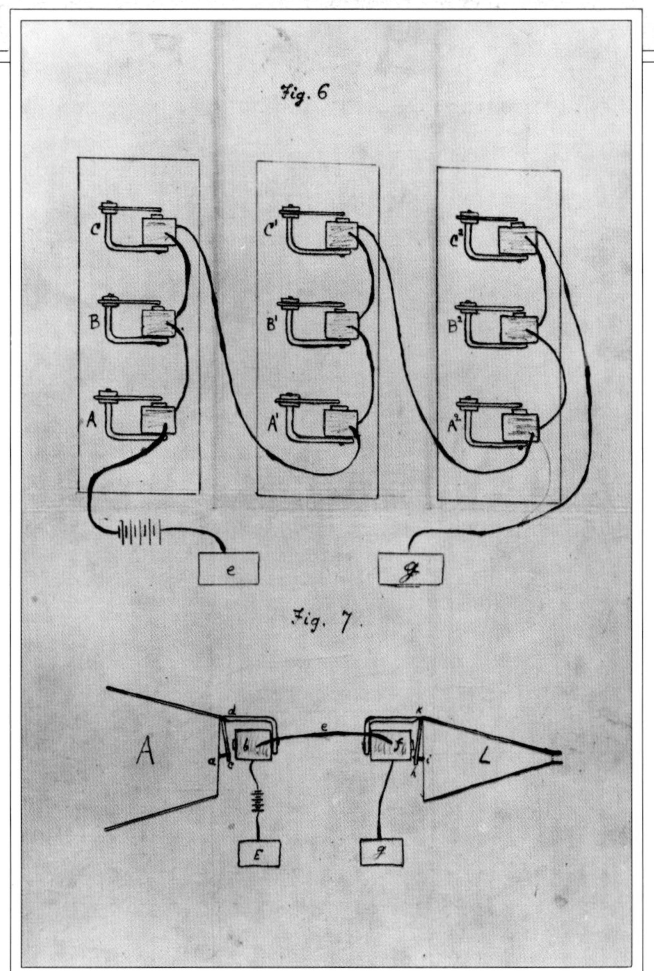

(Above) *The telephone circuit made simple, from mouth to ear: words flow through air, diaphragm, carbon granules, electric contacts, coils, and magnet.* (Below) *Bell's specifications for patent number 174,465.*

discussed the possibility of constructing one." It was an instrument Bell had been envisioning for a long time—in Boston, back in Brantford, wherever he was, whatever else he was working on. It was the answer to the question that had nagged at him since misreading Helmholtz years before: If vowel sounds could, indeed, be transmitted over a wire, well . . . why just vowel sounds? Why not all human speech? It was, in fact, the telephone.

TELEPHONE WASN'T A NEW WORD. THE GREEKS HAD USED it—*tele-phone* means *far-speaking*—possibly to describe the shouts from one Athenian hilltop to another. The Germans had used it; reference to a *telephon,* or a system involving megaphones, was first made in 1796. Sir Charles Wheatstone described his enchanted lyre as a *telephonic* in 1821. But none of them was talking about a talking machine. None, that is, until one C. G. Page of Salem, Massachusetts. In 1837, this unsung visionary put forth his theory of "galvanic music," which described the principle of magnetically induced electronic sound transmission over a wire.

In 1861, a German professor named Johann Philipp Reis showed up at the Physical Society of Frankfurt. There he demonstrated *his* telephone, which was a replica of the human ear, using a hollowed cork from a beer barrel as mouthpiece, a sausage skin stretched over it as diaphragm, and a violin case as resonator. It wasn't perfect, but it did transmit music over a 300-foot line. "From a purely technical standpoint, the telephone was born at the instant Reis first threw the switch," writes J. Edward Hyde in *The Phone Book,* "for his apparatus was able to receive and transmit sound—which is all a telephone is able to do."

There were other contenders in a field fraught with superstition and crossed wires. "The telephone was

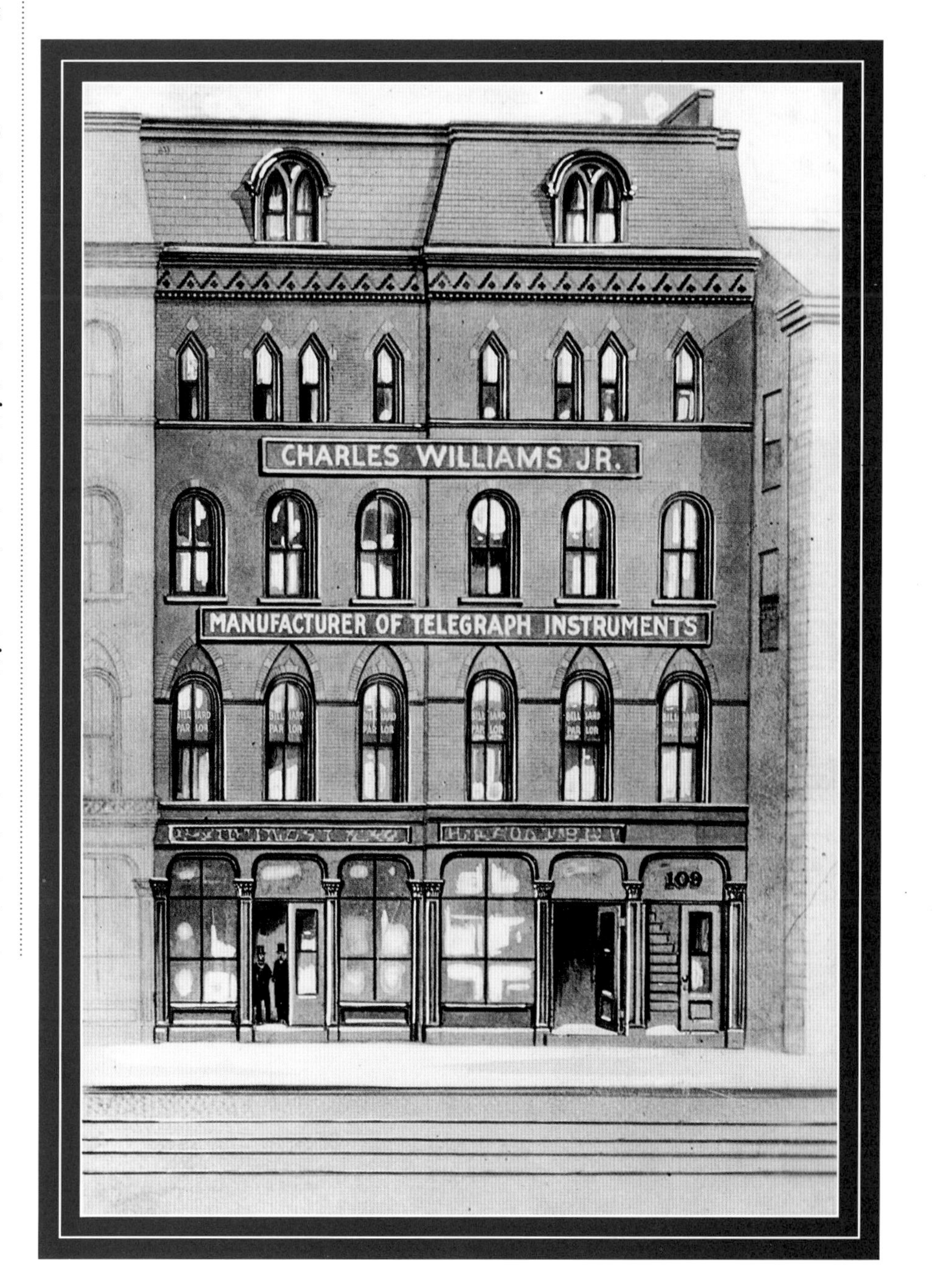

Birthplace: In the attic atop this Boston electrical shop, the telephone was born in June of 1875.

technically near—but philosophically far," explains John Brooks in his worthy history, *Telephone: The First Hundred Years*. "Human speech, as opposed to dot-and-dash code, was considered sacred, a gift of God beyond man's contrivance through science. Public reactions to the very idea of telephony in the 1860s and 1870s wavered between fear of the supernatural and ridicule of the impractical. People were made uneasy by the very notion. Hearing voices when there was no one there was looked upon as a manifestation of either mystical communion or insanity. Perhaps reacting to this climate, most physicists and electricians took it as an axiom that electricity could not carry the human voice. To have the freedom of mind to take the last step, there was needed a man whose thought was centered not on electricity but on the human voice, and the man was Alexander Graham Bell."

WATSON WAS ENTHRALLED WITH THE NOTION OF A speaking telephone. But Bell's angels, Sanders and Hubbard, were not. Their interest, and money, was in the harmonic telegraph, and they pressed Bell to get on with it. Valiantly, he and Watson persevered. Secretly, they pursued their telephonic dream. Toward both these ends, they explored the effects of metal diaphragms, vibrating strips, tuning mechanisms, fluctuating currents, magnetized reeds, and they even fiddled with the ear of a corpse.

And then, suddenly, when Watson was in the transmitting room and Bell in the receiving room—Eureka! "On the afternoon of June 2, 1875, we were hard at work on the same old job, testing some modification of the instruments," Watson recalled. "One of the transmitter springs I was attending to stopped vibrating and I plucked it to start it again. It didn't start and I kept on plucking it, when suddenly I heard a shout from Bell in the next room, and then out he came with a rush, demanding, 'What did you do then? Don't change anything. Let me see!' " What Watson did was a lucky accident. In trying to free a reed too tightly secured to the pole of its electromagnet, he had produced a *twang*.

This twang was more than just a simple sound. Because of its complex overtones and timbre, similar to those in the human voice, Bell was convinced that his vision of sending speech over a wire could actually work. This was it! The harmonic telegraph would have to wait. The next day, Watson built a crude machine—the first Bell telephone. That night, in the dark silence of Williams's deserted shop, the elated pair strung the world's first telephone wire. And the fifth-floor tones of Bell's voice—feeble, fuzzy, indistinct, but surely a voice nonetheless—were heard on the third.

The race was on. Telephonic fever was in the air. Even while Bell was frantically working to perfect the telephone, he was writing and drawing specifications to be filed with the United States Patent Office in Washington. His application was filed on February 14, 1876, and patent number 174,465—generally considered the most valuable patent ever issued—came through on March 7. Bell was twenty-nine. Three days later, the first telephone call came through. It was a call for help.

What made it possible was something called a liquid transmitter, which would deliver the current needed to transmit actual words. The liquid was battery acid. In another lucky accident, the liquid spilled—on Bell's britches. And it burned.

"Mr. Watson," he cried over the wire. "Come here! I want you!"

And Watson came running.

The reception and banquet given in honor of Elisha Gray, on the night of November 15, 1878, by his fellow residents of Highland Park, Illinois, will be long remembered, by citizens and strangers present, as an occasion of unusual enjoyment to all concerned.... It should be known that Elisha Gray, a resident of Highland Park, and a gentleman of superior scientific attainments, which have led him chiefly into the investigation of electrical subjects, is the individual to whom, beyond all doubt, the world is indebted for the original invention of the speaking and the musical telephone. It is not the intention to here enter into any discussion of this subject, which has, since the claim of another scientist was advanced, been taken to the cognizance of a judicial tribunal for the purpose of deciding the claim of priority of invention; but only that is stated which Dr. Gray's associates maintain, and which nearly all the scientists in the country concede, that to his brain is attributable the invention of that idea which has fairly worked a great revolution in telegraphy, and has demonstrated the startling capacities and the wonderful adaptability of electricity in that simple and yet almost marvelous instrument, the telephone.

—*A 1904 report published by the McRoy Clay Works on the Complimentary Reception and Banquet to Elisha Gray, Ph.D., Inventor of the Telephone*

As the speaking-telephone, in which magneto-electric currents were utilized for the transmission of speech and other kinds of sounds, was invented by me, I have described at some length my first instrument, and have also given explicit directions for making a speaking-telephone which I know, by trial, to be as efficient as any hitherto made.

—*Preface to* The Telephone: An Account of the Phenomena of Electricity, Magnetism, and Sound, As Involved In Its Action. with Directions for Making a Speaking Telephone *by Prof. A. E. Dolbear, 1877*

The two most prominent also-rans in the speaking telephone sweepstakes.

Events accelerated fast and furiously. It happened that 1876 was the year of the eagerly awaited Centennial Exhibition in Philadelphia. At the Fair, in extravagant surroundings, all the latest explorations in the fields of science, industry, and technology were on display. An international panel of judges marveled at Bell's machine. "My God, it talks!" exulted Dom Pedro, the emperor of Brazil, as Bell spouted Hamlet's soliloquy over the line from the main building one hundred yards away.

Bell and Watson took to the road, giving lectures and demonstrations with their Speaking and Singing Telephone. "Ahoy! Ahoy!" they'd greet each other over longer and longer distances, using the first official form of telephonic address (which Bell had initiated and would continue to use all his life).

On July 9, 1877, Bell and his patient patent partners, Hubbard and Sanders, founded the Bell Telephone Company (with 10 percent for Watson). Two days later, Bell married Hubbard's daughter, Mabel, and for a wedding present gave her his 30 percent of the company's shares. Off they sailed to England, where he introduced Queen Victoria to the telephone. Further events of the year included the first use of the telephone in newspaper reporting, the first telephone ad, and the first telephone made available for business use. It was Bell's father-in-law, Hubbard, in charge of early business affairs, who made a decision of major importance for the future of the company—that telephones would be leased, not sold, to subscribers.

Now the tremendously powerful Western Union Telegraph Company got into the act, better late than never. Having rejected the Bell Telephone Company's early invitation to purchase their patent for $100,000, the telegraph people realized they'd made a terrible and financially devastating mistake. Trying to make up for lost time, and having the wires and the wherewithal, they leapt into the fray, setting up a rival organization, The American Speaking Telephone Company. On the payroll, hustling to catch up with Bell, were two prominent inventors—Thomas A. Edison, working on a new kind of transmitter, and Elisha Gray, whose telephone Western Union had acquired.

Henry Fonda, Hollywood's Thomas Watson, demonstrates the early talking machine to a bewildered bystander.

Now nothing more than a footnote to history, Elisha Gray was an electrical genius from the Midwest who had, simultaneously with Bell, developed his own telephone device. Sadly for Mr. Gray, he filed his patent caveat on the very same day Bell did—but a few hours later. Victorious as he was with his other inventions—and victorious he was, managing to make over $5 million from them—he remained bitter to the end. Observed a Gray colleague somewhere along the way, "Of all the men who didn't invent the telephone, Gray was the nearest."

Bell's patent was being infringed on from all sides. Little Bell sued big Western Union and won. And so it went for many years, as over one hundred other telephone companies large and small forced the Bell Company into complicated legal entanglements. Bell and Watson, both of whom had retired from the company's day-to-day doings, were called upon to appear in courtroom after courtroom, all the way up to the Supreme Court, to defend their invention. In over six hundred lawsuits, their patent withstood attack, and they won every case.

The telephone was beginning to ring all over the country. In 1879, Western Union admitted defeat and pulled out. And then, with foresight and flair, Bell Telephone—now renamed the American Bell System—hired the shrewd Theodore Newton Vail away from his post as superintendent of the Railway Mail Service, as general manager of the company.

What American Bell needed most, Vail knew, was telephones. They were still being produced, very slowly, by the little Williams shop in Boston, and the public was impatient. So Vail decided to take advantage of the expertise and equipment at the Western Electric Manufacturing Company (a subsidiary of Western Union), which had, ironically, been established by Elisha Gray. By 1882, Western Electric was the exclusive supplier of Bell equipment and a part of the Bell System.

Thomas Edison Says "Hello!"

Tom was also responsible for the way we answer the telephone today. Originally people wound the phone with a crank, which rang a bell, and then said: "Are you there?" This took too much time for Edison. During one of the hundreds of tests made in his laboratory, he picked up the phone one day, twisted the crank and shouted: "Hello!" This became the way to answer the telephone all over America, and it still is.

—*Margaret Cousins,*

The Story of Thomas Alva Edison, *1965*

At the present time we have a perfect network of gas pipes and water pipes throughout our large cities. We have main pipes laid under the streets communicating by side pipes with the various dwellings, enabling the members to draw their supplies of gas and water from a common source.

In a similar manner it is conceivable that cables of telephone wires could be laid under ground, or suspended overhead, communicating by branch wires with private dwellings, counting houses, shops, manufactories, etc., uniting them through the main cable with a central office where the wire could be connected as desired, establishing direct communication between any two places in the city. Such a plan as this, though impracticable at the present moment, will, I firmly believe, be the outcome of the introduction of the telephone to the public. Not only so, but I believe in the future wires will unite the head offices of telephone companies in different cities, and a man in one part of the country may communicate by word of mouth with another in a distant place.

—A letter from Alexander Graham Bell to the organizers of the New Electric Telephone Company, March 25, 1878

Bell, 60, at his Nova Scotia estate, on a tetrahedral perch of his own design.

EVERY

MAN, WOMAN and CHILD

SHOULD CAREFULLY EXAMINE THE WORKINGS OF

PROF. BELL'S

Speaking and Singing Telephone,

In its practical work of conveying

INSTANTANEOUS COMMUNICATION BY DIRECT SOUND,

Giving the tones of the voice so that the person speaking can be recognized by the sound at the other end of the line.

The Sunday School of the

Old John Street M. E. Church,

Having secured a large number of Prof. A. G. Bell's TELEPHONES, will give an EXHIBITION at the CHURCH, 44 & 46 JOHN ST. N. Y. where all visitors desiring can make for themselves a practical investigation of the Telephone, by asking questions, hearing the answers to their questions, and listening to the singing conveyed through the Telephones from the other end of the line.

On Tuesday and Wednesday Afternoons

November 20th & 21st, 1877.

From 11½ A. M. until 7 P. M.

Admission to either Afternoon Exhibition 15 Cents.

AN ENTERTAINMENT

OF THE

Sunday School of Old John St. M.E.Church

WILL BE HELD

IN THE CHURCH,

TUESDAY EVENING, Nov. 20th, 1877, at 7.30 P. M.

CONSISTING OF

RECITATIONS by PROF'S SHANNON and McMULLEN,

SINGING by LITTLE NELLIE TERRY and others.

Concluding with the TELEPHONE EXHIBITION.

ADMISSION TO EVENING ENTERTAINMENT 25 Cents.

COME AND SEE THE TELEPHONE.

When the American Telephone and Telegraph Company was formed and incorporated on March 3, 1885, Vail became its president. By this time he had "established the basic framework of the Bell System as it exists today—vertically integrated supply; a network of licensees substantially owned by the parent company; emphasis on research and development; and strong supervision of the whole system by the parent," says telephone historian John Brooks. It was Vail's vision—"One Policy, One System, Universal Service"—that would be the company's credo for nearly a century.

THE SPINNERS OF MAGIC, THE MEN WHO HAD PUT WORDS on the wire, dropped the phone and went on to other things. Thomas Watson—infused with wanderlust and wonderlust—traveled extensively, studied geology and several languages, took up fencing, became a successful ship builder and a middle-aged Shakespearean actor, and died in 1934 at the age of eighty.

Bell and his beloved Mabel, to whom he was married for forty-five years, wintered in Washington and summered in Nova Scotia with their two daughters. He experimented with manned kites and hydrofoils, invented an electric probe and an audiometer, founded *Science* magazine and helped launch *National Geographic,* offered his home as a classroom for the first Montessori school in America, acquired twelve honorary doctorates, trifled with extrasensory perception, and never stopped dreaming. But no matter the excitement in the industry he had created, Bell forever refused to have a telephone in his study. He resented its persistent jangle. When he died of diabetes on August 2, 1922 at the age of seventy-eight, all telephone service throughout the United States stopped for one silent minute.

And yet with all he had accomplished, he still had one regret. "One would think I had never done anything worthwhile but the telephone," Alexander Graham Bell told Helen Keller as he spelled the words into her hand. "That is because it is a money-making invention. It is a pity so many people make money the criterion of success. I wish my experiences had resulted in enabling the deaf to speak with less difficulty. That would have made me truly happy."

Helen Keller (seated) with Bell, her lifelong friend, and Annie Sullivan, her teacher, in 1901.

To

ALEXANDER GRAHAM BELL

Who has taught the deaf to speak

and enabled the listening ear to hear

speech from the Atlantic to the Rockies,

I Dedicate

this Story of My Life

—Dedication in
Helen Keller's autobiography,
The Story of My Life,
1902

Just plugging

for a certain party to

Chapter 2

"Hello, Central!"

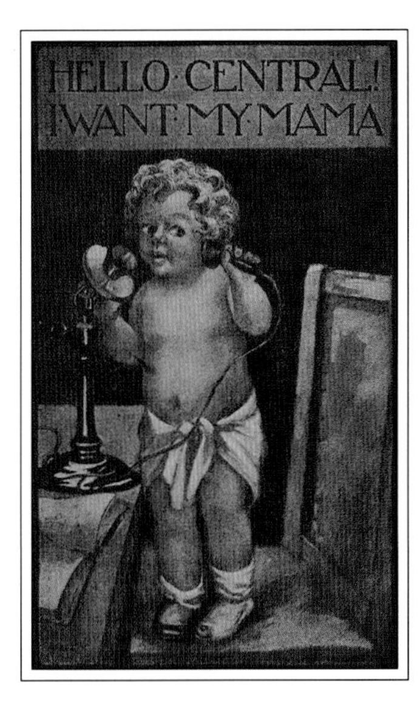

On the night of April 27, 1877, George Willard Coy was in the right place at the right time. As manager of the local Atlantic and Pacific Telegraph Company, he was in the audience at Skiff's Opera House in New Haven, Connecticut. He was there to hear a lecture-demonstration by Alexander Graham Bell. But he was more than a mere observer. It was over his telegraph wires, lent for the occasion and linked to the instruments themselves, that this first "dual" lecture was being made possible—between Bell in New Haven and Thomas Watson in Hartford.

Immensely impressed, the trailblazing Coy obtained from Bell the franchise to operate telephones in his city. Thus was the first telephone exchange—or central office—set up at 219 Chapel Street in New Haven, Connecticut, in January 1878. Within months, central exchanges run by telegraph office administrators and other local entrepreneurs would open in San Francisco, Albany, Lowell, St. Louis, Chicago, Keokuk, and Philadelphia. Within a few years,

(Above) *A turn-of-the-century postcard.* (Opposite) *A 1943 Hallmark card.* (Right) *"Weaver of Speech," from a Bell System ad, circa 1920.*

towns big and small from Montpelier to Monterey would be ready, willing, and wired.

What connected callers to one another were "switches" at a switchboard. In the first New Haven board, it was reported in *The Connecticut Home* magazine of January, 1938, "wire from discarded bustles was used, and in another switchboard placed in service in Meriden, a short time later, teapot cover handles and carriage bolts were successfully used to complete essential parts." The iron or copper wire connected customers' telephones to the telegraph wires, which connected to the central office.

The switchboard operator's primary job was to drag the wires across the bare floor and plug one subscriber into another. At first, these were young lads who had previously been employed at the telegraph office. "Two to four boys had to work together to complete a call, as they dashed from board to board to make connections. They also swept the floor, heaped coal on the fire, and collected bills from subscribers," writes Brenda Maddox in a 1976 study called "Women and the Switchboard."

But boys were ill-suited to the delicate work of telephony. Rowdy and restless, they took pleasure in insulting callers, pulling pranks, and crossing wires. "A

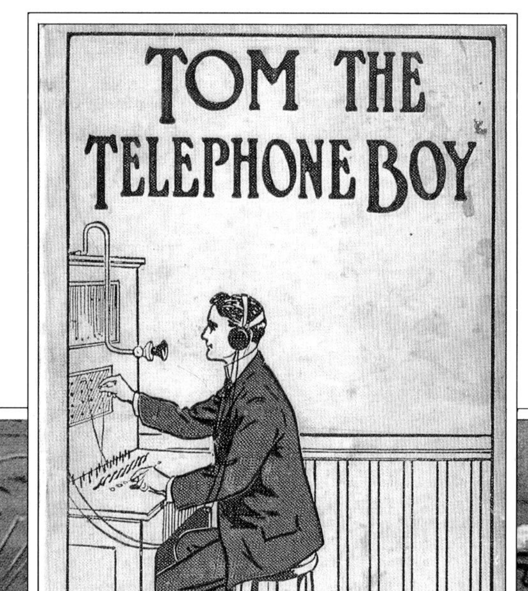

The first telephone operators: Boys, briefly.

perfect Bedlam" is how a Buffalo man described it. Young ladies, however, were known to be more patient, more composed, and certainly less costly.

And none was more patient or composed than Miss Emma M. Nutt, another graduate of the telegraph office. When, on September 1, 1878, she took her place at the central switching board of the Telephone Despatch Company in Boston, Miss Nutt became the first woman telephone operator in America. And the first to be addressed as "Central," a nickname that would stick for the next fifty years.

The Eve of her profession, she brought calm and discretion to the primitive board, setting the stage for many a respectable young woman eager to make a living as something other than schoolmarm, shop girl, or nurse—although the living was, at first, only ten dollars a month. The early operator was obedient, virtuous, and necessarily single. The unwritten rule was that she could not marry and would lose her job if she did. (In fact, the New England Telephone Company did not begin hiring married women until 1942.)

The operator's wardrobe was as prim and proper as she was. Starched and corseted, she perched stiffly on her stool, the metal headset flattening her upswept hair, the six-pound Gilliland Harness weighing down her shoulders and strapped around her waist. She wore

(Above) *Emma M. Nutt, the very first female telephone operator, hired in 1878.* (Right) *A disciple shoulders the Gilliland Harness.*

"Good Morning, The Plaza"

The Telephone Girl's Prayer

O Lord, for all I done to-day
To cause annoyance and delay
To make a person rant and rave,
For all wrong numbers I have gave,
And gave and gave when I'd be cryin'
For five three seven, thrrree seven, ni-yun,
For all the needless irritation
When I cut off a conversation,
The cusses—calls for information
Because of me—the slaps and slams,
The smashed receivers—darns and damns
I've caused this day—O Lord, for these
And all my sins,
Excuse it, Please!
Amen

—Oliver Herford

Beyond the splendor of the Palm Court and Oyster Bar at New York's Plaza Hotel is the Telephone Room, a rather unglamorous space hidden behind a carpeted door on the second floor, where twenty operators juggle an average of 560 calls an hour. Gone are yesterday's plugboards; today's consoles display a guest's name so he can be addressed by it. ("Good morning, Mr. Abernathy. Yes, I'm afraid it's still raining.") But not gone is the dress code. Despite their invisibility, women must wear hosiery, skirts, dresses. Men, jackets and ties and hair no lower than the collar.

When outside calls come in, operators will not ring a room unless given the name of the guest. Only in certain cases (royalty, presidents, movie stars) will they screen calls. They will not give out a room number or take a message without first ringing the room six times. Messages are written on paper and slipped under the door, but they are also typed on a computer so they can be seen on the TV screen (Channel 88) in the guest's room. Language barriers are almost unknown, thanks to the Plaza Hotel Language Directory, a list of employee names, departments, shifts, and languages spoken—from Arabic to Croatian to Hindi-Bengali.

Home Alone 2, released in 1992, caused a certain havoc in the Telephone Room, when two hundred to three hundred callers a day pretended to be Kevin McAllister trying to reserve a room; others merely asked for Kevin or Macaulay Culkin. In another year, a relentless child called all through Christmas Day asking for the fictional Eloise until the beleaguered operator put an end to it by pretending to *be* Eloise.

Otherwise, it's business as usual. "Good morning, the Plaza," they say, and then, "My pleasure." Composed under most circumstances, including bomb scares, they are, nonetheless, often blamed for forgetting wake-up calls because the guest has gone back to sleep. Models of patience, rarely ruffled, they answer dumb questions ("What flag is that hanging outside?") and alert security to a room that doesn't answer. And they remember every insult, especially when hurled from a guest as decidedly unamusing as Milton Berle. "We don't disconnect," sighs one, "but we would often like to."

a blouse with white linen collar and a long dark skirt. Focusing straight ahead at her switchboard, and never left or right, she was not allowed to cross her legs or suck a lozenge. She could not even blow her nose or wipe her brow without permission.

Such permission could be granted only by the austere female supervisor whose duties included watching, judging, preventing chatter, and plugging in to monitor performance. The first of these was Miss Katherine Schmitt. She joined up as an operator at the New York exchange in 1882 (or, as she called it, "the dark ages of the telephone" in her 1930 magazine memoir, "I Was Your Old 'Hello' Girl") and rose rapidly. It was the intimidating Supervisor Schmitt who instituted standard training methods, devised for clarity and diction to combat poor telephone transmission. Thus the number *ny-un* and "Hold the *ly-un, plee-izz*." And it was she who described the quintessential Bell belle: "The operator must now be made as nearly possible a paragon of perfection, a kind of human machine."

The early operator worked nine hours a day, six days a week, with no overtime. Burdened by the growing demands of the switchboard, she might burst into tears and flee to a designated retiring room and the comfort of a matron ready to massage her weary shoulders and send her back into the fray. A less emotional option was something called "standing relief."

Not until lunch time could she mingle with her female co-workers in the lunchroom provided by the company, since it was considered inappropriate for a young woman to travel about and wander into the exclusive domain of the all-male oyster bar. "I brought my lunch in a tin box," reported one dutiful worker. "It was not considered dignified to carry our lunch, so I tucked my tin box

in a black satin bag." Such was the carefully regulated life of a big-city company girl at the turn of the century.

In other quarters, matters were far less rigid. When Bell's original patent expired, the telephone business took on a chaotic ring. Any entrepreneur with a little capital and a get-rich-quick vision could get into the act and did. By 1900, there were six thousand new "independents"—telephone companies independent of Bell—privately owned and operated in every burg and hamlet. Thousands of country folk would, they hoped, be able to finally afford their own telephones. But cheaper equipment and a jerry-built office didn't necessarily mean a clean connection. Small-company service was haphazard and unreliable. Worse yet, because the Bell Company refused to interconnect with its unsanctioned offspring, a subscriber to one company

Fellowship, Loyalty, and Service: watchwords of the Telephone Pioneers, the world's largest industry-related volunteer group dedicated to serving others, founded in 1911.

I doubt that there is a more consistently courteous group in America than telephone operators.

—*Eleanor Roosevelt*

White House Phone Trivia

The first telephone was installed in the White House in 1878 during Rutherford B. Hayes' administration. President Hayes's first outgoing call was to Alexander Graham Bell, who was thirteen miles away. The president's first words: "Please speak more slowly."

☎

Teddy Roosevelt disliked the phone and used it only in an extreme emergency. Woodrow Wilson hated it and instructed the operators *not* to ring him. Herbert Hoover was the first president to have a phone on his desk.

☎

In 1933, Miss Louise Hachmeister became the first woman telephone operator in the White House. FDR called her his "hello girl" and his "phone detective" and gave her the nickname "Hacky." According to Congressman Claude Pepper, "Hacky could find anybody outside of the African Bush."

☎

Eisenhower was so accustomed to phone service from military and White House operators that he didn't know how to use a dial phone when he left the White House. He also persisted in calling operators "Central."

☎

Kennedy, more addicted to phones than previous presidents, overloaded the circuits and disrupted service several times. He said that the two things he would wish to take from the White House back into private life were *Air Force One* and the White House switchboard.

☎

Johnson used the phone more than any president in history. He was so fond of the female operators' voices that he would take some of them with him when he stayed at his ranch in Texas rather than deal with male military operators.

☎

President Reagan used a white, touch-tone call director with eighteen lines. Mrs. Reagan had her own private white touch-tone phone with one line.

—Mary Finch Hoyt, "Hello, This Is the White House." Good Housekeeping, *June 1986*

It has more than once occurred to me that, Long Island harboring as it does a number of lunatic asylums, certain of their more innocuous cases of arrested development are employed as operators. Certainly the operators give no indications to the contrary. In addition to the marked imbecility they have, which is even more annoying, a definite tendency to deafness which makes for any amount of fascinating mistakes. *Andrews* becomes *Vanderbilt;* a call put in for *Syosset* is miraculously answered at *Manhasset;* ask for a party in *Wading River* and the operator, as if showing an aversion to getting her feet wet, gives you the same number in *Riverhead.*

Not long ago I had occasion—in fact rather urgent occasion—to call the ferryhouse at Orient Point. The number is Greenport 51 and I dialed the operator. There is an art to making a Long Island operator answer the dial. It is a subtle art, akin to getting close to the beaver at work or luring the blue jay to feed from one's hand. For your Long Island operator is a creature of caprice and rarely if ever answers a first dial. She must be lured. Failing a moose horn or a crow call or even (and the impulse has seized me more than once) a shotgun, I usually find the best way is to stretch out comfortably on the bed, a good book in one hand, and with the other to keep up a patient rhythmic signaling on the receiver hook.

—Cornelia Otis Skinner, *Excuse It Please!*, 1936

could not telephone friends who subscribed to another. The wiring networks remained separate, and confusion was constant.

While the city operator was an anonymous voice with scripted words, her country cousin was anyone willing and able to handle the board and, most likely, someone everyone knew. She might be Daisy, the farmer's daughter, propped on a milking stool, flipping the switch for her father's independent exchange, or Miss Evelyn, whose omniscient position confirmed her status as the most powerful member of the community. Messenger and muse, eager to help by knowing everybody's business, she mediated marital mishaps and party-line disputes, reported the weather, announced train schedules and delays, pacified panicky children, shared recipes for blueberry crunch, and saved lives. "The worst night I ever endured was in April, 1921," recalled one valiant veteran. "This was during the Colorado flood, when I had to phone and warn all the subscribers on the west side of town that a 40-foot wall of water had been reported sweeping down from Pueblo."

IN THE BEGINNING, ALL TELEPHONE SUBSCRIBERS HAD names. Then they had numbers. The name-dropping began during a measles epidemic in 1879, when a family doctor in Lowell, Massachusetts, realized that the town's four operators were as vulnerable as everyone else. Thus did Dr. Moses Greeley Parker propose assigning a number to each subscriber. Then, in case the regular girls fell ill, a substitute operator who might be unfamiliar with people's names would still be able to make accurate connections.

By the new century, telephones had become so popular and operators so busy plugging and unplugging and misplugging that a more methodical approach was clearly needed. The most masterful solution was concocted and patented in 1891 by Kansas City undertaker Almon B. Strowger. The cantankerous mortician had, for years, been obsessed with the notion that Central was plugging his potential customers into the competitive parlor. He was determined to devise a "girlless, cussless telephone."

"Strowger came up with a system that could serve ninety-nine telephones, based on a sort of windshield wiper in the central office that automatically moved around to touch the contacts of the number being called when the caller pressed the correct number of times on two buttons attached to his telephone," explains historian John Brooks. "To get num-

There are about seven or eight phrases that you use and that's it: "Good morning, may I help you?" "Operator, may I help you?" "Good afternoon." "Good evening." "What number did you want?" "Would you repeat that again?" "I have a collect call for you from so-and-so, will you accept the charge?" "It'll be a dollar twenty cents." That's all you can say.

A big thing is not to talk with a customer. If he's upset, you can't say more than "I'm sorry you've been having trouble." If you

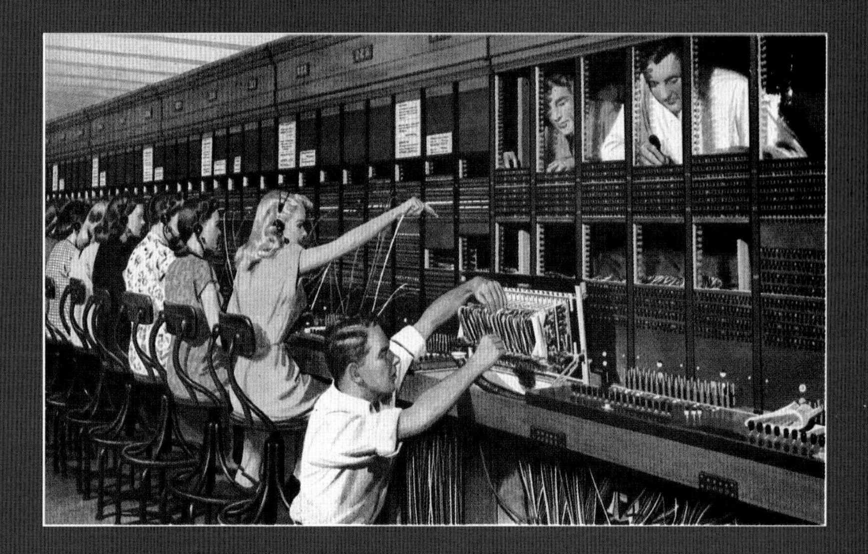

get caught talking with a customer, that's one mark against you. You can't help but want to talk to them if they're in trouble or if they're just feeling bad or something. For me it's a great temptation to say, "Gee, what's the matter?" You don't feel like you're really that much helping people.

The girls sit very close. She would be not even five or six inches away from me. The big thing is elbows, especially if she's left-handed. That's why we have so many colds in the winter, you're so close. If one person has a cold, the whole office has a cold. It's very catchy.

You try to keep your fingernails short because they break. If you go to plug in, your fingernail goes. You try to wear your hair simple. It's not good to have your hair on top of your head. The women don't really come to work if they've just had their hair done. The headset flattens it.

Your arms don't really get tired, your mouth gets tired. It's strange, but you get tired of talking, 'cause you talk constantly for six hours without a break.

—Heather Lamb, Telephone Operator,
in Working, *by Studs Terkel, 1972*

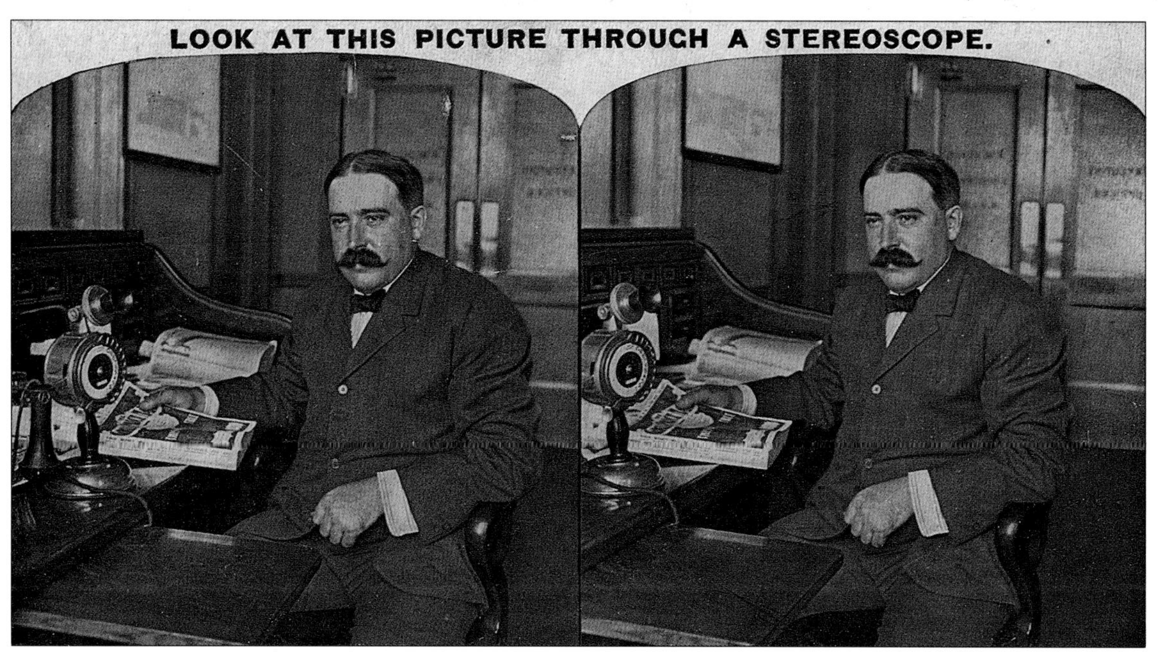

Mr. R. W. Sears, president of Sears, Roebuck & Co., proudly displays his Strowger dial phone.

and would become even more so with the advent of Long Distance.

The first domestic long-distance call, between Boston and Providence, had been made in 1881. North, south, east, and west, poles and wires crisscrossed the country. On January 25, 1915, a grand ceremony marked the first coast-to-coast

ber 99, the caller had to push each button nine times. This remarkable contrivance of a man who might have been expected to be handier with embalming fluid than with mechanical devices was, beyond question, the first successful automatic switch-board, and the lineal forebear of the 'step-by-step' system, the original dial-telephone switchboard."

The Strowger automatic switch came in, but, to his dismay, the operator didn't go out. In fact, although the electrical impulse significantly began replacing the spoken word, and the customer began, in a crude way, to dial numbers by himself, the operator was still a major player. She was still crucial in supplying Information,

T*elephone switchboard. Buzzing in darkness. We see* AUNTIE MAME *wrestling with the plugs. Several are already in place.*

AUNTIE MAME: Widdicombe, Gutterman, Applewhite, Bibberman, and Black—good morning. One moment, I'll connect you with Mr. Gutterman. *(She plugs.)* Widdicombe, Gutterman, Applewhite, Bibberman, and Black, good—Yes, Mr. Bibberman, I'll get you Mr. Applewhite. *(She crosses two plugs, one from each board to the other.)* Widdicombe, Gutterman, Applewhite, Bibberman, and Black. Good morning. Long Distance—*(She plugs.)* Mr. Widdicombe, I have your San Francisco call. *(She plugs.)* Yes, Mr. Bibberman. *(Innocently.)* Oh, did I give you Mr. Gutterman instead of Mr. Applewhite? I'm sorry, Mr. Bibbercombe—uh, Bibbe-bib. *(She pulls down a plug from one side and starts to put it in the other side. But there is already a plug in the hole she wants to put it in. She pulls down the offending plug from the hole with her other hand and holds it up, addressing it as if it were a person.)* Mr. Applewhite, what were you doing in that hole with Mr. Gutterman? *(She now has two plugs, one in each hand. Forgetting where they are supposed to go, she thrusts one plug in her bodice to free her hand for another call. She puts up another plug.)* Mr. Widdicombe, I'll try to reconnect you with San Francisco. Now, let me see . . . *(She starts rearranging the plugs at random, sticking one plug in her mouth.)* Mr. Bibberbip is in there and Mr. Gutterwipe is talking to—And where's Mr. Applewhite? *(The board is going crazy now. She is plugging and unplugging frantically now. Buzz.)* Oh, there you are, Mr. Applewhite. Yes, sir. *(She dials.)* Eldorado five—two one two one. Yes, sir. Hold on, Mr. Widdicombe. I'll find San Francisco. *(Buzz.)* Widdicombe, Gutterman, Applewhite. Oh! Supervisor? I did? *(She pushes a key.)* I'm afraid you gave me a wrong number, sir. There is no Applewhite five, Mr. Eldorado. *(The lights fade as the frantic buzzing continues.)* I don't know what I did with San Francisco.

(The lights are out.)

—*Jerome Lawrence and Robert E. Lee,*
Auntie Mame, *1957*

Rosalind Russell gets her wires crossed as Auntie Mame of stage and screen

I've been in switchboard work since the days of the party line. This middle finger and I have always served you, working ourselves to the nub. *Look.* My ear is cauliflowered from years of having my headset hooked into it. I've got switchboard hump from leaning forward and dialing, dialing, dialing . . . shoulder bursitis from plugging, unplugging, plugging, unplugging . . .

—Lily Tomlin as Ernestine, 1993

Manual System

Mary has a thingamajig clamped on her ears
And sits all day taking plugs out and sticking
plugs in.
Flashes and flashes—voices and voices calling
for ears to pour words in
Faces at the ends of wires asking for other
faces at the ends of other wires:
All day taking plugs out and sticking plugs in,
Mary has a thingamajig clamped on her ears.

—*Carl Sandburg*

The "Hello" Boy

In 1969, I worked for one day as an AT&T Directory Assistance (nee Information) operator, in Brooklyn. I had been with AT&T for some two and a half years in the legal department as a junior attorney. At the time, AT&T was being threatened with a strike which would idle non-supervisory workers, and I was junior enough to be tabbed for strike duty to keep the system going regardless. Sure enough, the workers went out, and I was told that on the following day I was to proceed to a windowless little blockhouse of a building in Brooklyn. Despite a Bronx upbringing that did not prepare me for the vagaries of Brooklyn geography, I found my way to my new duty station. I was quickly taken in tow by an amiable but firm chief operator—surely an appropriate analogue for the army's master sergeant stereotype—who explained what was expected of me.

I spent the day from nine to six waiting for a little white light to appear, at which time I pushed a button, answered, "Information. May I help you?" and was immediately electronically plugged in to those of the fair borough needing assistance in obtaining phone numbers in order to reach others within the five boroughs. I soon learned that the operator is called upon for more than just getting a number to go with a name. Thus, people had no problem in asking for the pizzeria nearest to the intersection of Hoyt and Schermerhorn streets (places formerly as foreign to me as Trafalgar Square or Tianenmen Square), or where they could see Elizabeth Taylor's latest opus. And through diligent study of directories and other nonlegal research, I found I really could help them—and I loved it. In fact, I found that, as an operator, I really had my ear on the personal lives of my interrogators and their families. The fact is that people simply do not consider operators more than an extension of a lifeless telephone network, and, while I was preparing to respond to their requests, I was privy to both arcane family secrets and extremely creative swearing. Perhaps my greatest rewards were the exclamations like: "Hey, you're a guy. You're doin' a helluva job, and [prophetically] the phone company ought to hire more of you fellas."

In retrospect, I look back on this experience as perhaps the single most satisfying day (the strike was soon settled) in a quarter-century career at AT&T. I really got things done (today, we'd say I had closure, not the inconclusiveness of ten-year cases); I really dealt with real people; I really felt like I was helping my community—I was a cog in the great public service machine that was the Bell System. Since then, I've had important assignments: running the legal side of divestiture; creating my own law department of over 300 people in the new AT&T; dealing with billion-dollar cases. But that one day toiling in the sweet vineyards of a Brooklyn telephone facility told me a lot about me and the company I worked for.

—Alfred A. Green, attorney, 1993

In 1929, when I was about fourteen and lived in Kansas, one of my close friends, Ruth, who was a couple of years older, often ran the telephone switchboard. Telephone exchanges were privately owned in those days, usually by a local resident. Ruth's father was the town electrician and he owned the telephone company in our town as well. Often on a dull Sunday afternoon, some of my friends and I would walk downtown, climb the stairs to the telephone office above the drugstore and spend a fascinating few hours with Ruth while she operated the switchboard. When some titillating conversation was going on, we would be especially quiet so Ruth could listen and hopefully repeat what she heard.

The most exciting events when we visited the telephone office were the general alarms. Sometimes during summer storms, lightning strikes would ignite prairie grasslands or hit a barn or home. Fires spread rapidly with hot summer winds to fan them. Ruth would plug in all her circuits and a series of rings would alert people in the area. Very soon volunteers (town residents) would come running from every direction toward the fire house where a hose cart was kept. We had a clear view from the front windows of the telephone office and could watch as they hooked up the cart, attached it to someone's truck, and hurried away to the fire. The flour mill would also blow its whistle to alert the townspeople if the situation was quite serious.

—Ruth Ware Koelzer, 1993

telephone call between the elderly Alexander Graham Bell in New York and the ever-faithful Thomas Watson in San Francisco. In 1927, as Charles Lindbergh crossed the Atlantic, so did the first overseas telephone call.

Once, the operator was known as Central, or the telephonist, or the telephone girl, or the "Hello" girl, or the Voice with a Smile. Then she became "O" for Operator, for Omniscient, for On call. She is still there, but "Information" has become "Directory Assistance" and rapidly disappears upon connecting the inquisitive caller to an electronic voice. Supervisor Schmitt's early vision of a "human machine" has taken on new meaning. The operator's job is now reduced to such asocial acts as scanning the board, throwing switches, reversing charges, refunding quarters from coin phones, giving credit for wrong numbers, and checking to see that a busy line is really busy. A "human machine" indeed.

1878
1882
1892
1897
1902
1905
1914
1920
1928
Today
"How many of 'em can YOU remember?"
"Some of the old-timers must look pretty strange to you. But not to me . . . I *made* all of them.
"I started making telephone apparatus of all sorts in 1877 . . . did such a good job that I was asked to join the Bell Telephone team 'way back in 1882.
"Telephone users get more and better service for their money in this country than anywhere else in the world. I've helped to make this possible by efficient *manufacturing* of uniform, high quality equipment .. by volume *purchasing* of all manner of supplies for the Bell Telephone Companies . . . by *distributing* to them, through my warehouses in 29 principal cities, the telephone equipment I make and the supplies I buy...by skillful *installation* of central office equipment.
"That is a huge job . . . especially now when the demand for telephone service is at an all-time peak.
"Remember my name . . . it's Western Electric."

Chapter 3

Form and Function

(Above) *The first commercial telephone, which went into service in 1877.*

The Williams shop throbbed with energy as a small band of electrical experts, directed by Thomas Watson, tested metallic diaphragms, electromagnets, and iron membranes in their haste to turn out the world's first consumer telephone. The initial result looked something like a large Brownie camera. A lens-like opening at one end was used for speaking *and* listening, which meant that the user had to heave this monstrous marvel back and forth from mouth to ear. Boxy with a wood finish, the eagerly awaited instrument was introduced in 1877. A Boston banker was the first customer to lease a pair of telephones (it took two to make a conversation) from the newly organized Bell Telephone Company. With a private line strung from his downtown office to his house in Somerville, he could now let the family know exactly when he would arrive for dinner.

Within months, there were three thousand telephones out and about—phones manufactured and leased exclusively by Bell, the only legal provider—and their form and function were continually evolving. The Butterstamp, so called because the receiver looked like the butter stamp found in every farmer's kitchen, was churned out in 1878.

(Opposite) *A Bell telephone ad from the forties.* (Right) *Mae West on a golden oldie.*

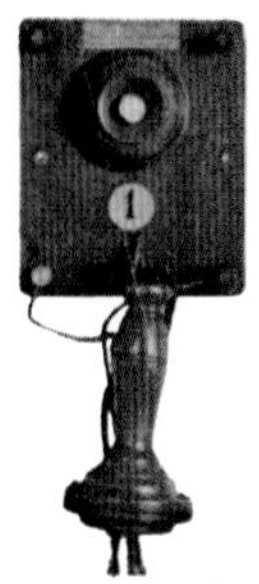

1878 Butterstamp. A combined receiver-transmitter required the user of this set to talk or listen separately, in sequence.

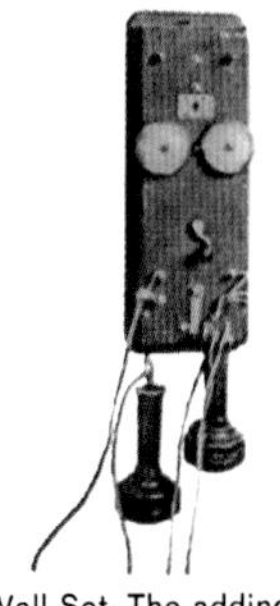

1878 Wall Set. The adding of a second receiver-transmitter made it possible for one to talk and listen at the same time.

1880 Blake. Invented by Francis Blake, Jr., this new set was able to transmit the voice with greatly increased clarity.

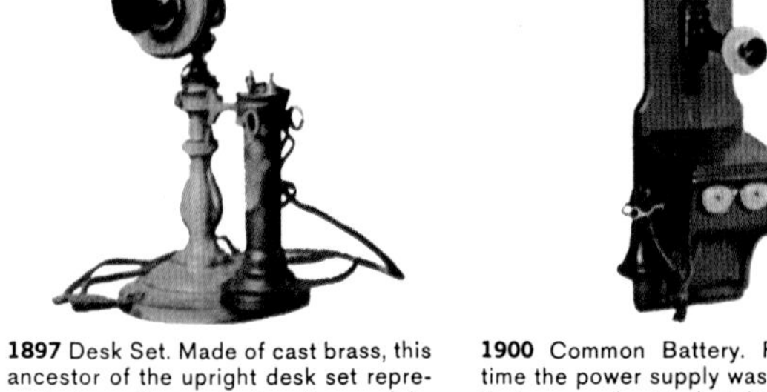

1897 Desk Set. Made of cast brass, this ancestor of the upright desk set represented a refinement of earlier models.

1900 Common Battery. For the first time the power supply was taken out of the home and put in the telephone office.

1907 Magneto Wall Set. A more modern version of the phone with a built-in generator for signalling the operator.

1928 Desk Set. This combined receiver and transmitter and became popularly known in the U. S. as the "French phone."

1930 Desk Set. Its new oval base and color finishes distinguished this later Desk Set from the earlier 1928 version.

1937 "300" Type Desk Set. Engineering innovation in desk set design: bell was placed in the base of this model.

It was a great advance over the early box because the ear-and-mouthpiece (it was still one) could be lifted from the transmitter unit fixed to the wall. But customers still had a clumsy time switching the same transmitter-receiver from mouth to ear. The well-to-do finessed the problem by acquiring two instruments—one for talking and one for listening. A button signaled the operator.

Turning a crank, not pushing a button, summoned the operator with the Wall Set, introduced later that year. This version incorporated the two transmitters for talking and listening. But sound quality was so poor, so cluttered with clicks and buzzes and whirrs, that parties at both ends of the line had to shout to be heard. People who could afford such a newfangled implement expected fewer wrinkles.

Every tinkerer with a workshop was trying to refine the thing. In Menlo Park, New Jersey, Thomas Alva Edison—who had already invented a vote-recording telegraph system and an electric pen—was developing something called a carbon transmitter to improve voice trans-

1949 "500" Type Desk Set. Most widely used set in the U. S. today. Major new feature was volume control for bell.

1954 "500" Type Color Desk Set. Basic colors—white, beige, green, pink and blue—made phones fit modern decor.

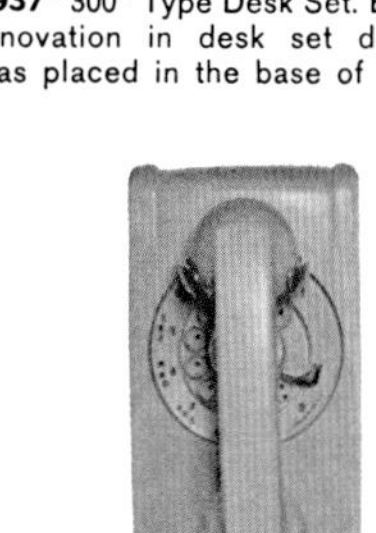

1956 Wall Telephone. A convenient extension phone in many basic colors for kitchen, basement, patio and garage.

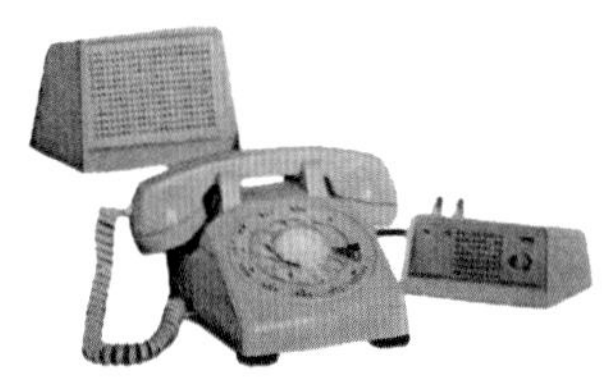

1958 Speakerphone Set. Many features for business offices including hands-free conversation, group conferences.

1958 Call Director® Telephone. Multi-purpose business phone, equipped with up to 30 pushbuttons for handling calls.

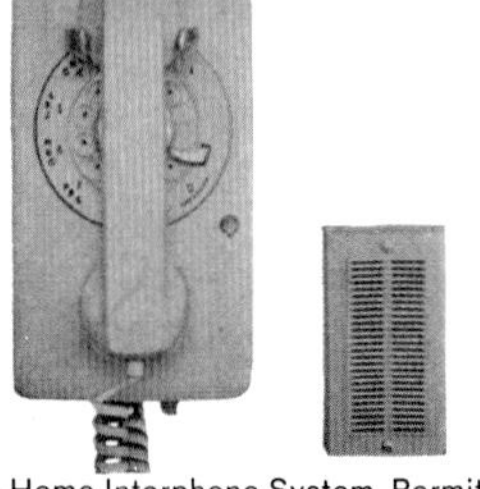

1960 Home Interphone System. Permits communication between rooms of home, in addition to usual telephone service.

1962 Panel Phone. This wall telephone has a flush-mounted base, retractable cord, and comes in many smart colors.

1964 Touch-Tone® Telephone. A new era in telephoning—pushbuttons instead of a dial. Makes calling easier, faster.

1965 Trimline®. A space-saving dial on the handset, with a dial light, provides increased convenience to subscribers.

PICTUREPHONE. New dimension in telephoning. You hear and see one another. Attended service now in some cities.

mission, as was a German immigrant named Emile Berliner in Washington. Edison was the real threat to the Bell Company because he was working for Western Union, the competitor with whom Bell was embroiled in continuous legal battles over patent infringements. At the same time, in Weston, Massachusetts, Francis Blake, Jr., was fiddling with something called a variable-resistance transmitter. And by 1879, a Bell telephone set incorporating the Berliner-Blake principles—which now made a conversation less a shouting match than it had been—was mounted on a decorative stand and perched on the most sophisticated desks in town.

Despite the speedy increase of mechanical improvements, customers were experiencing supremely irritating technical difficulties. Electricity—in the form of the lightbulb, the trolley, and even the thunderstorm—was

(Left) *Stan Laurel deals with the new-fangled dial.* (Opposite) *A promotional poster from Western Electric.*

Take the old stand-up Bell model of 1914. . . . Form never followed function more faithfully—and, if the functionalists of design are right, the telephone of 1914 ought now to be sitting with the immortals of art. It lasted a long while, to be sure (officially until 1927 and in out-of-the-way places long after that). It bespoke its usage honestly. Probably no one ever started talking into its earpiece by mistake, and that is more than can be said for some of its successors. With its ringing apparatus hung separately on the wall, however, and its angular combination of iron foot, post, mouthpiece, hook, and earpiece reflecting the piecemeal procedure of its engineers, the telephone of 1914 remained throughout its active career the artless little brother of the flivver and the biplane.

—*Wallace S. Baldinger,* The Visual Arts, *1960*

It is hazardous to stand at the foot of a pole while a lineman is working above, ascending or descending. Warn all persons, especially children, to keep away for the following reasons:

(1) He may drop tools.
(2) He may dislodge splinters and chips.
(3) A lineman's gaff may cut out.

A second lineman, preparing to ascend, should always wait until the first man has reached the working position and placed his safety strap. In descending, one lineman should remain in his work area until the other has reached the ground. *If possible on poles which are wet or on which there is snow or ice, climb with the gaffs engaging the slippery side of the pole in order that the hands may engage the drier side and reduce the hazard of slipping.*

Pins, crossarm braces, insulators and hardware other than pole steps do not furnish a safe support as they may pull loose or break. The gloved hands may be cut on such devices that may be rough or broken. Do not use this equipment for support by the hands or for attachment of the safety strap.

—The Use and Care of Pole Climbing Equipment: A Field Manual
Prepared by Accident Prevention Committee,
Edison Electric Institute, New York

causing major interference along the telephone wires. "Such a jangle of meaningless noises had never been heard by human ears," noted Herbert N. Casson in his 1910 *History of the Telephone.* "There were spluttering and bubbling, jerking and rasping, whistling and screaming. There were the rustling of leaves, the croaking of frogs, the hissing of steam, the flapping of birds' wings. There were clicks from telegraph wires, scraps of talk from other telephones and curious little squeals that were unlike any known sound. The lines running east and west were noisier than the lines running north and south. The night was noisier than the day, and at the ghostly hour of midnight the babble was at its height."

The more telephones, the more telephone wires. The more telephone wires, the more telephone poles. By the late 1880s, the urban landscape had become a forest of towering poles with multiple crossarms carry-

Snow imperils phone service in New York City during the blizzard of 1888.

Lines to a Lineman

No word of pen or stroke of artist's hand
No flowered phrase or oratory's boast
Need tell the story of the world you've made.
'Tis writ upon the pages of the land
From north to south—from coast to coast.

Those poles you mount
Those lengthened strands you string
Are not just sturdy uprights in the sky
That march across the miles in proud parade.
You've made them into words that help and sing
A doctor's call, good news, a lover's sigh.

Deep etched in time the record of your skill
The work you've done—your willingness to do
The fires and storms you've tackled unafraid,
Your signature is carved on every hill
Yours, too, the creed—"The message must go through."

—*Bell Telephone System*

Under a Telephone Pole

I am a copper wire slung in the air,
Slim against the sun I make not even a clear line of
shadow.
Night and day I keep singing—humming and
thrumming:
It is love and war and money; it is the fighting and
the tears, the work
and want,
Death and laughter of men and women passing
through me, carrier of
your speech,
In the rain and the wet dripping, in the dawn and
the shine drying,
A copper wire.

—*Carl Sandburg*

ing so many wires that the sky was nearly obliterated. In the country, birds found a new perch upon which to rest and chirp. And elsewhere throughout America, "They blend along small-town streets/ Like a race of giants that have faded into mere mythology," John Updike poetized in "Telephone Poles" nearly a century later.

THE OAK MAGNETO WALL SET WAS DEVISED TO DEAL with the continuing cacophony by combining the Blake transmitter with Bell's receiver. It became a household fixture around 1882 and remained the standard wall model until the late 1890s. It was the first telephone from Western Electric, by now the designated manufacturer and supplier for the Bell System. The crank was relocated from the front to the right side. The ringing apparatus occupied a box of its own, while visible "cow bells," "sleigh bells," or "tea cups" merrily jangled

to announce a call. (Before bells, there had been squeals.) A second box below held the electrical works: the Leclanche wet battery, sitting in an acid-filled jar with the potential to dry up, overflow, or explode.

But a dignified turn-of-the-century tycoon could hardly be bothered with a telephone on the wall. He needed something near at hand. Various desk sets, or deskstands, appeared through the 1890s; the swankiest was the brass model introduced in 1897. Later models, popular into the late twenties, would be known as the Candlestick and the Upright.

There were at least 250,000 telephones in use by now, and no matter how the design was evolving, the

device itself still depended upon a fallible human being down at Central. The solution—that primitive dialing system devised by Almon Strowger, the cranky undertaker—turned out to be the forefather of all dials.

Thus came a mad race to place a dial on every telephone. The first was placed on the standing desk set, but it was a delicate balancing act, and the instrument tended to tip. The solution: the "cradle" desk phone. Because this design concept—a single, hand-held piece that combined the speaking and listening apparatus—had originated in France, people called it the French phone. Squat, black, oval-based, and made of nickel-plated brass, it sat more solidly on the telephone table. Its receiver-transmitter, cradled on the instrument instead of hooked to its side, was an upgraded version of the headset previously used only by telephone linemen.

Since the telephone was company property and

(Above) *"Always in reach, yet never in the way" bragged the 1920s brochure for the Dormeyer Extension Bracket, which lifted phone from desk.*

provided rather than chosen, it could hardly have been expected to harmonize with the furnishings found in upper-class boudoirs and living rooms. It was an industrial object, designed for maximum efficiency, not beauty. And, whatever its style, it remained leashed to a clunky bell box.

In 1930, the Bell Laboratories invited ten artists and craftsmen to enter a redesign competition. "Some people weren't quite sure where to put [telephones]," recalled prolific industrial designer Henry Dreyfuss, writing of the challenge in his 1955 memoir, *Design for People.* "They were sometimes kept inside plaster globes of the world or cabinets or dolls with fluffy skirts. Because placement had a bearing on design, we had to determine what people did with phones.

"It was flattering to be included in such a group, and the prospect of a thousand dollars was attractive," he went on. "But I suggested that a telephone's appearance should be developed from the inside out, not merely created as a mold into which the engineers would eventually squeeze the mechanism, and this would require collaboration with Bell technicians." Dreyfuss did not participate in the competition—which, in fact, nobody won because the entries proved unsatisfactory to the telephone brass—but he was hardly out of the picture. Later commissioned to work up some designs, he did—and they became classics.

Dreyfuss's "300" desk set series came out in 1937. The ten-finger holes were clearly marked with letters

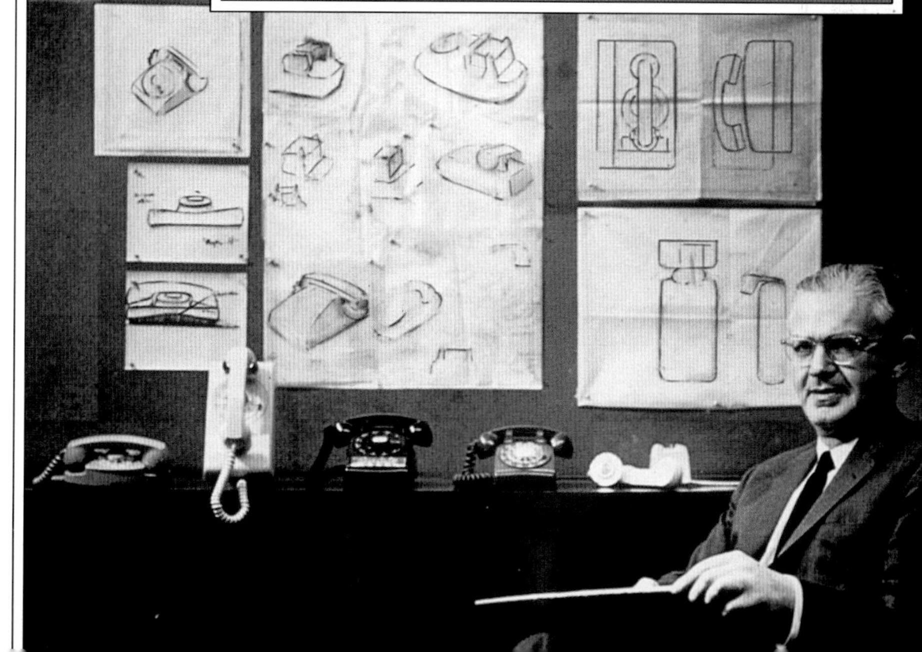

(Above, left) A finger-saving plastic dialer; (above right) *the famous Model 300 telephone;* (center) *its cousin, the 500, and* (below) *their designer, Henry Dreyfuss, shown in his New York office.*

Highly personal...

in red and numbers in black and seemed welcoming to a hand now involved in the everyday act of dialing. Clickety-click went the turning dial. The compact body was made of heavy phenolic molding; the bell was enclosed in a square base. Glamour—in the form of white or gold—could be found only on the movie screen. "Like the Model T," observes Phil Patton in *Made in U.S.A.*, "the standard model 300 came in any color you wanted so long as it was black."

Drefuss's "500" telephone, issued in 1949, was the very model of postwar modernity. Low-slung and accessible, it hugged the table like a Studebaker hugged the road. Its cord was coiled, its volume adjustable, its classic body an engineering feat of black plastic outside and new materials inside. The dial was expanded from three inches to four and one-quarter inches, and the letters and numbers—white characters molded in black—were placed outside the clear-Lucite finger wheel.

it's

fun

to phone

The fifties were a time of "turning out fantasy on an assembly line," as Thomas Hine put it in *Populuxe*, his celebration of that synthetic era. "Color and styling were applied to objects that had always been viewed as purely practical. The advertising industry was able to associate goods with moods, and many consumers fell into a pattern of buying a car, or even a stove, not because the old one was worn out, but simply because they felt like it." In 1954 consumers could not yet buy their telephones, but they could, for the first time, choose from a limited line of decorator colors.

The Princess phone, introduced in 1959, was a majestic scheme to entice the female segment of the market: housewives and their teenage daughters. Pro-

duced in pastels, this slim item featured "graceful styling" and "lovely lines." It nestled in niches and glowed in the dark.

In 1963, Bell Labs came up with a better mousetrap: Touch-Tone service. To access this, a phone needed push buttons. To get push buttons, a customer needed a new phone.

And then a newer phone. In the sixties and seventies, America was presented with the Speakerphone, Data-Phone, Centrex, the Trimline, the Circle, the Touch-a-Matic, a red-white-and-blue replica of the ancient Candlestick, a creamy version of the Hollywood-inspired Celebrity, and the one-piece, lightweight, giraffe-like Ericofon (designed to eliminate the repairman's perpetual nightmare, ROH—or receiver-off-the-hook). With the breakup of AT&T, in the 1980s, it finally became possible to buy—and no longer lease—a phone of one's own, and the choices became infinite. Novelty phones took the shape of pocketbooks and high heels, ketchup and Coke bottles, elephants and lips. Mobile phones hit the road and the beach, the golf course and the jet plane. In fact, there have been over one thousand styles designed and dialed since Bell's first bulky box.

(Left) *The statuesque Ericofon, developed in the 1950s by L. M. Ericsson of Sweden and manufactured in Ohio by the North Company, sat on its dial and switch button.* (Right) *Don Adams as TV's inept Maxwell Smart.*

JUNE 27 1916

I AM sending you this to let you know that the New York Telephone Company (Bell System) has installed a telephone in my residence at

206 Market St

My telephone number is

5056

c/o The BRESLIN

C 2044

Chapter 4

The Telephone Book

PEnnsylvania 6, BUtterfield 8, and PLaza 9 are telephone numbers that resonate with a mingling of memory and desire. But before they were these evocative exchanges, telephone numbers were just numbers. And before they were numbers, they were names.

The first name in phone book history was Reverend John E. Todd of New Haven, Connecticut. The good man had responded to a four-page prospectus distributed by George Willard Coy and partners, who were trying to attract customers in order to set up the world's first central telephone office. Rev. Todd was willing to pay $4.50 per quarter for the service, but one customer alone wasn't enough. So Coy hired some canvassers to go door to door. With the addition of twenty more daring townsfolk, the doors opened to the New Haven District Telephone Company at 219 Chapel Street.

Three weeks later, on the frosty morning of February 21, 1878, the world's first directory—a single

LIST OF SUBSCRIBERS.

New Haven District Telephone Company.

OFFICE 219 CHAPEL STREET.

February 21, 1878.

Residences.

Rev. JOHN E. TODD.
J. B. CARRINGTON.
H. B. BIGELOW.
C. W. SCRANTON.
GEORGE W. COY.
G. L. FERRIS.
H. P. FROST.
M. F. TYLER.
I. H. BROMLEY.
GEO. E. THOMPSON.
WALTER LEWIS.

Physicians.

Dr. E. L. R. THOMPSON.
Dr. A. E. WINCHELL.
Dr. C. S. THOMSON, Fair Haven.

Dentists.

Dr. E. S. GAYLORD.
Dr. R. F. BURWELL.

Miscellaneous.

REGISTER PUBLISHING CO.
POLICE OFFICE.
POST OFFICE.
MERCANTILE CLUB.
QUINNIPIAC CLUB.
F. V. McDONALD, Yale News.
SMEDLEY BROS. & CO.
M. F. TYLER, Law Chambers.

Stores, Factories, &c.

O. A. DORMAN.
STONE & CHIDSEY.
NEW HAVEN FLOUR CO. State St.
" " " " Cong. ave.
" " " " Grand St.
" " " Fair Haven.
ENGLISH & MERSICK.
New Haven FOLDING CHAIR CO.
H. HOOKER & CO.
W. A. ENSIGN & SON.
H. B. BIGELOW & CO.
C. COWLES & CO.
C. S. MERSICK & CO.
SPENCER & MATTHEWS.
PAUL ROESSLER.
E. S. WHEELER & CO.
ROLLING MILL CO.
APOTHECARIES HALL.
E. A. GESSNER.
AMERICAN TEA CO.

Meat & Fish Markets.

W. H. HITCHINGS, City Market.
GEO. E. LUM, " "
A. FOOTE & CO.
STRONG, HART & CO.

Hack and Boarding Stables.

CRUTTENDEN & CARTER.
BARKER & RANSOM.

Office open from 6 A. M. to 2 A. M.
After March 1st, this Office will be open all night.

sheet of fifty names printed on white card stock—was delivered to subscribers. It wasn't alphabetical, and there were no numbers. Among the eleven residence customers were Coy, his partners, and their lawyer. Then there were the Physicians; Dentists; Meat and Fish Markets; Hack and Boarding Stables; Stores, Factories, Etc., and Miscellaneous (which included the Police Office, Post Office, *Yale News*, and Quinnipiac Club).

Almost simultaneously, Boston's Telephone Despatch Company issued a directory, businesses only. The list was alphabetized and categorized, from agricultural to woolens; subscribers numbered sixty-seven. The single sheet proclaimed: "The above Company proposes, and is now prepared, to establish direct TELEPHONIC COMMUNICATION between each and every business house in this City. Whatever inquiry you wish to make, business you wish to transact, or message you

The first telephone directory.

In New York, some 37 percent of NYNEX customers—that's more than one in three—pay $1.95 a month to keep their names, numbers and addresses out of print. The $23.40 per annum goes for more than the cost of disappearing ink. It buys you virtual assurance that no one can obtain your number no matter what hard luck stories they tell. The only time a number is given out is when it's "in the interest of public safety." In other words, a number can be subpoenaed or acquired when proper documentation is presented.

Otherwise, no go. The directory assistant doesn't even see your number on his terminal. Which means that if your mother rings Information claiming a family emergency, she's switched to a supervisor who has access to another computer system. That supervisor will size up the situation and, if convinced, will call *you* to say that Mother is at her wit's end.

There are cities across the country—and more and more of them—equipped with something known as E (for enhanced) 911. In these up-to-date locations a 911 call is accompanied by visuals that include the caller's phone number and address. The idea is that if distraught people are unable to give that info themselves, it's still available.

Incidentally, the official term is not "unlisted number" but "nonpublished number."

—David Finkle, writer, 1993

TRafalgar-6

I could reach my father from anywhere by calling a TRafalgar-6 number. My father was a TRafalgar-6 guy; my girlfriend Susan was an ENright-9 girl. I had become an ACademy-2 guy, having taken an apartment near Columbia University. The other day from a taxi window I saw Susan, now in her mid-forties, crossing the street, looking good. I thought, ENright-9. That triggered TRafalgar-6. That brought up my father, a certain gesture of his—the circling of his right index finger to make a particular kind of point. But now, to any generation younger than my own, ENright-9 is 369, and TRafalgar is 876 or 877 or 874 or 873—numbers unable to bring fathers to life in taxis.

I've called my friend David from all over the world. He's 744. The other night, I asked him if he knew he had a RHinelander exchange. He said, "What do you mean?" David is 35.

I work at a radio station at 42nd Street and Third Avenue: 986-7000. Not long ago, it occurred to me that I was the single employee out of 73 who knew the station's telephone number to be YUkon 6-7000. The one other who would have known was recently terminated and has retreated to his 421, a designation without the history of language. These empty codes are creeping in for keeps: 633, 691, 249 (my daughter, alas), and others lacking in the poetry that was once New York.

I mourn the words we've lost. Think of them: CHickering, WAdsworth, SChuyler, GReenwich, ORegon, COlumbus.

You could learn about a fella by knowing his exchange. A MOnument fella was up near 100th Street and West End Avenue. You could picture him coming downtown on the IRT, strolling first to 96th and Broadway for the newspapers, passing the Riviera and Riverside movie theaters (both gone). The ATwater girl was an East Side girl, a taxi-hailing girl, on her way to her job at Benton & Bowles. A CIrcle fella was a midtown fella, entering his CIrcle-7 Carnegie-area office with a sandwich from the Stage Deli. And what about a SPring girl, a SPring-7 girl, twirling the ends of her long brown hair as she lay on her bed talking to you on the phone? A Greenwich Village girl. A 777 girl is nothing. She is invisible. She is without irony, seldom listens to music.

John O'Hara did not use the word BUtterfield frivolously. PEnnsylvania-6-5000 still dances to its forties tune—you can hear the guys in the band shouting it out. PLAza. There were so many. My friend Joanna is PLaza-2. I think of the word "PLaza" when I see her at Christmas. My publisher is PLAza-1, not 751. Bertha and Larry, my father's best friends. They both died a few years ago, burying their PLaza-8 with them.

It's a losing situation. It is global and elephantine. Gone from Los Angeles is CRescent. Gone from Boston are COpley and KEnmore. Gone from London are KENsington and OXford and PICcadilly. Gone from Paris are ODEon and CARnot. Phone language, intimate stuff, lost to the digital world of now, to the oblivious finger pecking out numbers rat-a-tat-tat, as you read this, rat-a-tat-tat 20766067444521778842.

Last week, without fuss or fanfare, I called my florist to send flowers to my wife. It was her forty-first birthday. I dialed CHelsea. CHelsea-3.

—*Jonathan Schwartz,* New York *magazine, December 21–28, 1987*

"That Is an Unlisted Number"

The people who are listed in the telephone directory—and don't misunderstand me, they're the backbone of the nation—resent the people whose numbers are not listed. Reticence about one's telephone number, they feel, is an act of social or professional striving, a symptom of a swelled head. Sometimes it is, no doubt; but the whole unlisted number question isn't that simple. The dividing line between listed and unlisted is, as the political economists say, vertical rather than horizontal, definitely not confined to any one class. Beyond that, it is difficult to make any hard and fast rule.

I have observed that the very rich are usually listed; so are doctors, unsuccessful actors, book makers, and others whose careers depend more or less on telephone calls. Those who are only moderately rich tend to have unlisted numbers, probably because they have butlers but no private secretaries and are in some danger of being reached by somebody in the outside world and talked into an unwise investment or an overlarge donation. Successful actors have private numbers, and so do people who write pieces for magazines. . . .

After a year an unlisted number becomes a liability—if it was ever an asset. An unlisted friend of mine, a man with a statistical turn of mind, figured out recently that his number was in the possession of eleven girls he didn't like any more, forty-three other people he had *never* liked, a former business associate who was suing him at the moment, a discarded masseur, three upholsterers who had made estimates on re-covering a sofa, a political group with which he was no longer in sympathy, a dealer who had sold him a car in '41 and was trying to buy it back, and an unidentified alcoholic who never called until after my friend's bedtime. He'd like to start life over again with a listing in the directory, he says, but he hasn't the courage to face the consequent social disgrace.

—Russell Maloney, *Atlantic Monthly*, May 1944

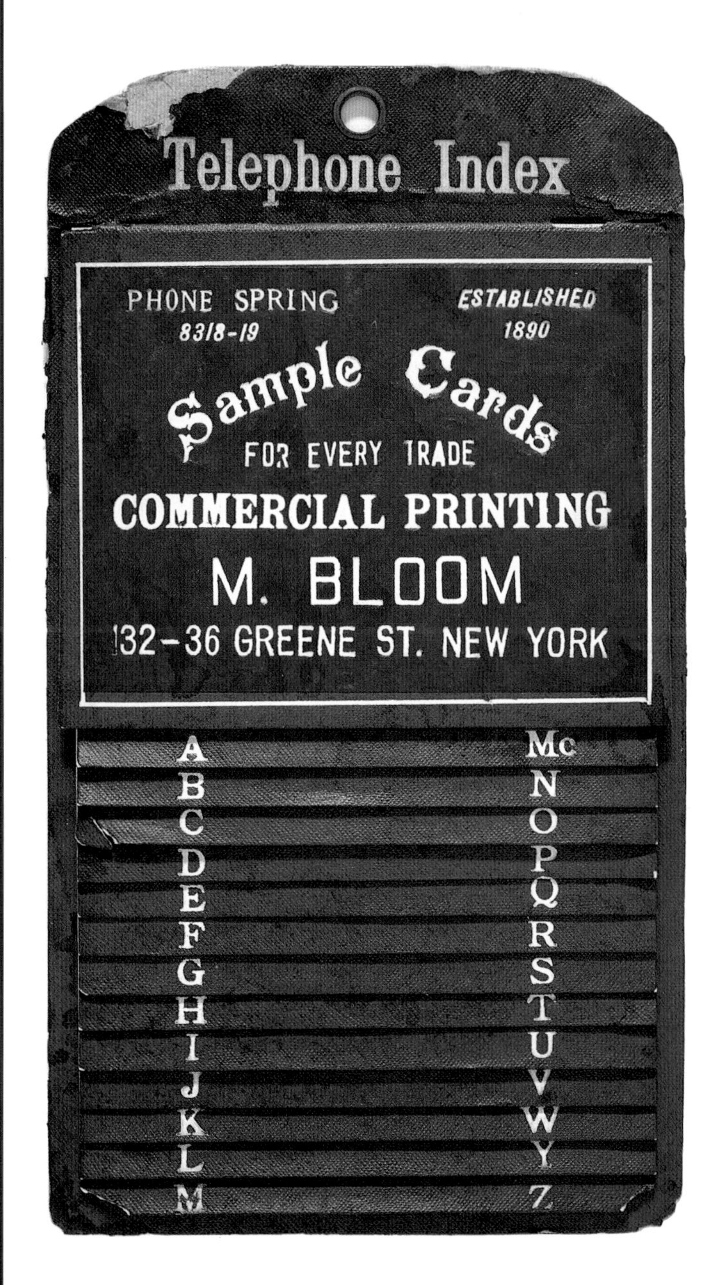

wish to communicate with any other business house or profession in the City, can be done while seated at your desk. This is done by communicating through the Telephone, to a Central Office, that you desire to speak to Mr. A, B, or C. In an instant your communication is made direct and complete, and you can carry on your conversation,—which it is *impossible* for a third party to

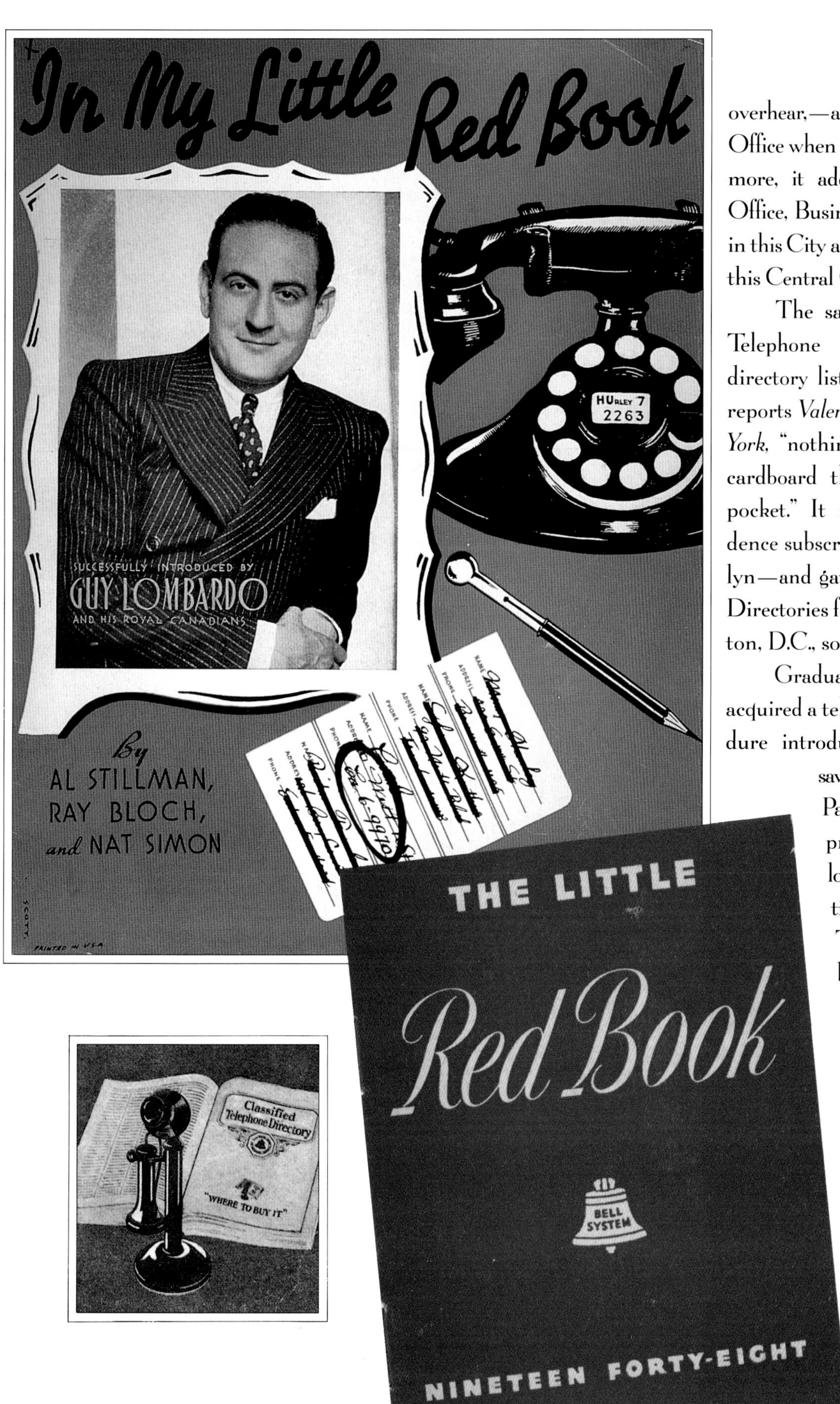

overhear,—again signaling the Central Office when you are through." Furthermore, it adds sternly, "Every Bank, Office, Business House and Profession in this City are expected to connect with this Central Office System."

The same year, the New York Telephone Company published a directory listing 252 names. "It was," reports *Valentine's Manual of Old New York*, "nothing more than a piece of cardboard that would fit in a vest pocket." It included seventeen residence subscribers, with five in Brooklyn—and gave their addresses as well. Directories for Chicago and Washington, D.C., soon followed.

Gradually, each subscriber acquired a telephone number (a procedure introduced in 1879 when the savvy Dr. Moses Greeley Parker of Massachusetts prescribed it). Listings got longer and longer. Pages multiplied and grew in size. The book got bulkier and bulkier. Binding became inevitable. A cover came next.

Thus did the telephone book evolve.

In most regions of the country, the pages were white. Until some

There are, at the moment, over two hundred Yellow Pages publishers in the United States—collating, proofing, printing, binding, distributing over 350 million directories across the country, and generating over $9.3 billion in annual revenue. While our fingers do the walking, the Yellow Pages are galloping.

A "business" subscriber to the telephone company (as opposed to a plain, everyday "residential" subscriber) is entitled to a White Pages listing and a one-line Yellow Pages listing. For more exposure, however, he can pay for a listing in boldface, or an in-column ad (which contains no graphics but can use color and measures from a 1/2 inch to 2-1/2 inches), or a trade ad (which can proclaim his logo). But most exposure is found in a display ad, which can run from 1/16 of a page to a full page.

Not every ad makes the book, however. There are ethics at work here. Superlatives are taboo. Photos are unwelcome; they don't print well. Then there are the "trouble ads," which might be embarrassing due to copy or art—or proximity. ("Abattoir" and "Abortion" would be an unfortunate juxtaposition; "Abortion Alternatives" and "Abortion Providers" must never appear on the same page.) NYNEX Directories, which publishes the Yellow Pages in New York and New England (as well as in Gibraltar and the Czech Republic, oddly enough), has a warning list for its layout artists.

"All Pages Containing These Specific Headings Are To Be Considered Sensitive Pages!!" the caveat reads. Headings to be heeded include Cemeteries, Entertainers, Escort Services, Funeral Directors, Marriage Brokers, Massage, Mobile Home Parks, Pregnancy Information Services, Sanitariums, Singing Telegrams, Social & Human Services (the latter a cause of frequent concern, especially in these politically correct times, since it can include anything from Elderly Persons Services to Gay Lesbian & Bisexual Organizations to the Ku Klux Klan).

Of the Yellow Pages' headings, which change according to fashion—Boston has had its Chicken Pot Pies, the 1970s its Pet Rocks, today its Lead Paint Detection & Removal Services—Physicians & Surgeons remain, year in and year out, in sickness and in health, the most popular.

of them turned yellow. Why yellow? Telephone Company legend has it that in 1883, a printer in Cheyenne, Wyoming, simply ran out of white paper and plucked from the nearest ream—which just happened to be yellow. Customers liked the new hue. It was easy to locate, and it was easy to read.

As the cities—and their telephoners—grew, the Central offices became more and more overworked, more and more inadequate. Branch-office exchanges sprang up hither and yon, using a name-number system. These names—"Main," "Eastern," "Hollywood," "Bensonhurst"—designated their locations. Later names were borrowed from local trees and landmarks, international heroes and grand old families.

Telephone technology advanced beyond the early Strowger system, and by 1921 the rotary dial was at hand. First came a wheel with ten finger holes, a number in each. This simple system, with its limited numerical combinations, was sufficient to serve small towns. Then, thanks to William G. Blauvelt of the American Telephone and Telegraph Company, the dial incorporated numbers and letters. This

A Book Review

Manhattan Telephone Directory, 1970–71
New York Telephone Co. paperbound, 1,896 pp. Free.

Book collectors would do well to stock up on the 1970–71 Manhattan Telephone Directory. Doubtless it will figure among the finest editions of this work, surpassing all previous editions in layout, typography and literary style.

The introductory pages are a farrago of typography—light and bold, upper and lower case—a subtle symbolism evoking the pluralistic tone of things to come. The occasional use of Ultra Bodoni italics has an implicit ethnic appeal, and sounds a typographic clarion call whenever the reader's attention may be flagging. Page 21, a thin scrim, is profligate with white space as it leads us gently into the body of the work. This page provides the authors with the undisturbed chance to display the imprimatur of the patron, whose current trademark is a tightly muscular, somewhat Teutonic version of a former, more Arabesque, symbol. Finally, the myriad characters of the opus are arranged in the new computer-adaptation of 6-point Bell Gothic. No listing of the characters ever goes beyond the width of a tidy, readable 12 picas.

Those who have casually brushed aside earlier efforts by the anonymous authors as superficial—"all characters and no plot"—would do well to reread the current work, for it is filled with surprises, fresh viewpoints and valuable services at no cost. . . .

On the international-plot front, the reader will quickly identify the protagonist America on page 67, and the antagonist China on page 319. The selection of only one character with these names suggests a pitting of two world forces presently engaged in a period of accommodation. Symbolically interspersed with the two world giants are 21 Frances and thirteen Englands.

On the domestic front, the authors have avoided coming to grips with the issue of polarization and have achieved a moderate, middle-of-the-road position by defining about 1,400 Whites, 400 Blacks and over a thousand Browns. While racial confrontation still looms large in this book, the authors have made it quite clear that the balance of power is held by Middle America's silent majority, for there are a total of 5,700 Smiths and Joneses.

The handling of the law-and-order issue will please this silent majority but raise the hackles of the radiclib establishment. In the book's over-simplified optimistic analysis, for example, there are more than 50 Laws and twelve Justices but only one Crime and no Riots or Unrests. Optimism goes even further in this edition, with four references to Money but only one to Tax. Good cheer and a positive outlook are also found in one Loss and eleven Wons.

It is to the credit of the Directory's authors that substantial attention has been given to environment and ecology. Devotees of the Zero Population Growth movement will take heart to know that there is only one Birth and no Babies. Conservationists will be delighted to find America's favorite, the Pine, referred to 50 times, but only slight concern with Maple, Oak and Elm. The national forest question has obviously taken precedence over Manhattan's local environment, for while the trees which abound on the city's streets are the plane and the sycamore, the first gets only one mention and the last is completely ignored.

Acknowledging the continuous appeal for cookbooks, the authors have included this market in the present edition. While it would take too much space to deal adequately with the overall culinary richness of the book and the proliferation of Sugars, Coffees, Teas, Butters, Bourguignons, Tetrazzinis, Salmons and Bolognas, the author's taste for herbs and spices reflects quite conventional interests, for the most frequently called for are Pepper and Salt, with Dill, Parsley, Mustard and Thyme casually tossed in. The increasing international sophistication of cooking arts may require that the next edition include an Oregano and Coriander, for which there are presently no listings. There are, however, Peaches, Oranges, Apples, and, yes, we have no Bananas. Music enthusiasts as well as gourmets may be interested to know there are thirteen Partridges and two Pear trees.

—Sol Chaneles and Jerome Snyder, New York *magazine, March 22, 1971*

> "Where's the suburban phone book?" I asked after pulling out all the books tucked under the telephone table.
>
> "What?"
>
> "The suburban phone book. I want to call Short Hills."
>
> "That skinny book? What, I gotta clutter my house with that, I never use it?"
>
> "Where is it?"
>
> "Under the dresser where the leg came off."
>
> "For God's sake," I said.
>
> "Call information better. You'll go yanking around there, you'll mess up my drawers. Don't bother me, you see your uncle'll be home soon. I haven't even fed *you* yet."
>
> —*Philip Roth,* Goodbye, Columbus, *1959*

system provided manifold possibilities, essential in densely populated and increasingly talkative cities. Blauvelt's letter-number plan used an edited alphabet (no Q or Z) so that in combination with numbers, the caller could now dial the named exchanges himself, unassisted by the operator. She wasn't obsolete, though. Zero, or "O," on the dial still summoned her for Long Distance.

Letter-number combinations varied from place to place, year to year. A seven-digit combination became standard after World War II. Every telephone wore identification in the form of a number card. The Candlestick model wore it like a hat,

In America, everyone asks me why my telephone number is in the directory. To a mere Englishman this is a bewildering question. Over there, everyone's number is listed. Once, when I looked for the number of the Westminster Libraries, I noticed, above it, the words "Westminster, Catherine, Duchess of." If she can have her number in the book, so can I.

When Mr. Letterman told the world my number was listed, for about six months afterward I did receive half a dozen calls threatening my life. That seemed a small price to pay. In England, I received six such calls a day. However, I still didn't change my number.

If your number is not ascertainable, you will be stuck with just your friends. What good is that? The whole point of having a telephone is that you can talk to all the world without feeling compelled to reconstruct your appearance (which in my case takes about three hours).

For this reason, the telephone is my favorite musical instrument and the only one that I can play. The key to success with the telephone is to answer it cozily. So many people say, "Yes. What is it?" in such a brusque manner that the caller immediately thinks he has chosen a bad moment. In his panic, he may even ring off.

Sad to say, I may soon be denied this delightful form of communication because I cannot use the now universal "woodpecker's" instruments: I need a dial. Everything else is science fiction to me.

—*Quentin Crisp, author and actor, 1993*

Abbvtr

Having heard that Mr. Russell W. Templin is the man who has charge of, and in many cases has dreamed up, the given-name, street-name, locality, and business-category abbreviations in the New York telephone books, we called on him at his office for a semasiological chat. "We abbreviate in an effort to keep all listings down to one line," he told us. "One million copies of the Manhattan directory are published twice a year, and one extra line, in a million copies, uses up as much paper as two copies of the directory. We eliminated periods, commas, and apostrophes long ago, and we've been cutting, cutting, cutting, right down through the years. We're the only people who don't put u's after q's—we have antiqs, laqrs, and liqrs." Mr. Templin, an earnest man with an abbreviated build and severely abbreviated hair, did not spell these abbreviations out. He pronounced them, with admirable clarity, looking as though he were about to swallow his teeth. He continued, "We take liberties with addresses—Rockfelr Plz, GrndConc, Chrlton, Flus, and so forth—which the post office certainly frowns on for purposes of superscription, but we don't approve of being used as a mailing list."

The NY Tel Co is a seasoned public utility, and Mr. Templin, who joined it as a supervisor of its directories in the nineteen-twenties, has identified himself cozily with its storied past. "I was an industrial consultant in leather goods when they offered me a job," he told us. "I realized at once that the leather business was ephemeral and that abbreviations would span the decades—had already done so, as a matter of fact. We had Bros, Co, Cor, and the abbreviated connective in the very first directory, in 1878, and we used to put res after a name. Later we cut it down to a simple r, but for some years the matter of residence has been liquidated entirely, as a wasteful extravagance. We used to abbreviate some street names and list others in full, and we're still dogged by inconsistencies, as a result of weaving in new listing material with the old without disturbing the old. I daresay you might still discover an r here and there in the directory, after one of our older subscribers, and even a lawyer or two spelled out in full." Templin is the father of lwyr and atty, as well as of elk, tlr, lds tlr, bty sln, undtkr, physthrpst, pltry, lthr, flwrs, fthrs, prfums, precs stns, ventlatg, clng, and whol flrst. He has no idea how many abbreviations he is in charge of. He generally asks permission from subscribers to shorten their professions, and never uses a Patk or Fredk without getting an O.K. from the parties affected. "Playing with a man's name is playing with fire," he said, with the air of a man who had been singed. "During the war, a number of commissioned officers asked to be listed with their rank. We considered this at some length and finally decided to adopt standard Army and Navy usage, except that no matter how many stars a general has, he is only a genl to us. I must say it cost us a lot of paper. We try not to be arbitrary, even when a fellow insists on having his street name spelled out. Queens is the worst place for street names. We've developed listings like 113-11 JamAv RichHl and 46-10 GrntpAv Wdsd, but even so, Queens runs into more two-line listings than any other borough." Mr. Templin's most vexing current abbreviation problem is monsignor, which he cut down to mgr several years ago. A number of monsignors and managers besought clarification, and he changed it to mnsr. "It's just come back again to my desk today," he said. "We probably cut it a little bit too fine."

Templin or no Templin, a few items have got the Telephone Company completely buffaloed: "We simply don't know what to do with honey, vanilla, shoulder pads, mushrooms, psychologists, and psychoanalysts," he said, "so we leave them alone." With this stmnt, he concluded our interview.

—*The "Talk of the Town,"* The New Yorker, *August 3, 1946*

affixed to the instrument above the mouthpiece. On desk sets, the small, circular card settled in the middle of the dial. It showed the subscriber's phone number, with an added letter—J, M, R, or W—for party liners. When the area code was introduced in the early fifties, it, too, was added.

Words were on the way out. For one thing, letter combinations were limited. For another, the complex spelling of some exchange names led to confusion. And for another, all-number dialing made international phoning possible since most foreign cities never had alphanumeric dial plates anyway. And so, city by city, beginning in 1961, Bell Telephone phased out exchange-name dialing. Seventeen years later, to the country's dismay, the WAlnuts, LOcusts, SPruces, and MAgnolias were just so much dead wood.

MEANWHILE, BACK AT THE BINDERY, THE PHONE BOOK was gaining weight. By the thirties, streamlining was in order. By removing the letter "r," which denoted residence, reducing the space between words, and changing the type face, pages could be eliminated. In such a way did the 1938 Philadelphia book, for instance, redesign its columns and unload 248 pages.

With so many alphabetical and classified listings, some sort of separation was inevitable. It came sporadically. In large cities, phone companies started producing twin books: the White Pages and the Yellow Pages. The first telephone directory had six business headings. Today's average Yellow Pages might have five thousand. The Patent Solicitors and Salt Casters of yesteryear have been replaced by Computers-Dlrs. and Asbestos Abatement & Removal Svces. *O tempora! O mores!*

I'm sorry I'm late but I was trying to call a friend of mine in Philadelphia with the new simplified dialing system—it's really wonderful—I dialed 211 and asked for this hotel in Philadelphia—211 told me to call information—that is 555-1212, but first I had to dial 411 to get the code number for Philadelphia—which is 215. I dialed 215-555-1212 and found out the hotel's number was 215-341-2345. I was calling from 212-682-6789 and my credit card number was 301-245-3578. 211 dialed 215-341-2345 calling from 212-682-6789 charged to 301-245-3578 and a voice came on and said, "Congratulations, that's the winning number for the Irish Sweepstakes!"

—The opener for Joan Rivers's nightclub act in the early 1960s, dealing with "the brand-new area-code telephone dialing system," she explains, "allowing America to dial anywhere—and at that time driving everybody crazy."

"Just 30 minutes
AND MY LUNCHEON'S
ALL ARRANGED"

Chapter 5

"Give Me a Ring"

At first, the telephone was a toy of the upper class. What's more, it was daring. Complicated. Intimidating, as anything electrical was. Expensive (the leasing fee was $150 a year in New York and $100 in Chicago, Philadelphia, or Boston—a substantial amount of money in the 1880s). And it was owned by them, not you. "Telephones are rented only to persons of good breeding and refinement," an early advertisement reminded potential customers.

Not everyone had a telephone in those days, nor did everyone want one. "If Bell had invented a muffler or a gag, he would have done a real service," grumbled Mark Twain when his first telephone was installed in 1878. "Here we have been hollering 'Shut up' to our neighbors for centuries, and now you

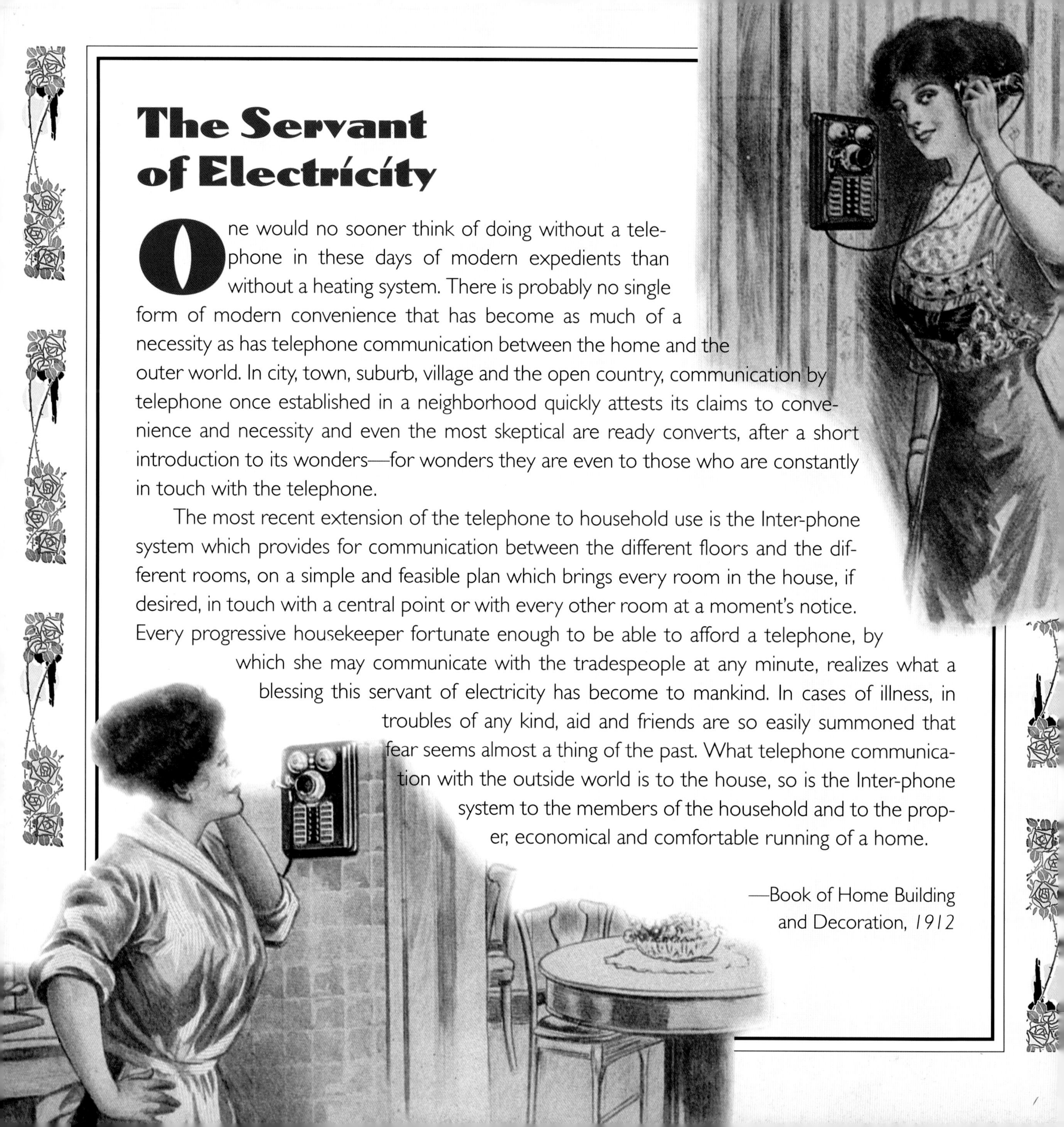

The Servant of Electricity

One would no sooner think of doing without a telephone in these days of modern expedients than without a heating system. There is probably no single form of modern convenience that has become as much of a necessity as has telephone communication between the home and the outer world. In city, town, suburb, village and the open country, communication by telephone once established in a neighborhood quickly attests its claims to convenience and necessity and even the most skeptical are ready converts, after a short introduction to its wonders—for wonders they are even to those who are constantly in touch with the telephone.

The most recent extension of the telephone to household use is the Inter-phone system which provides for communication between the different floors and the different rooms, on a simple and feasible plan which brings every room in the house, if desired, in touch with a central point or with every other room at a moment's notice. Every progressive housekeeper fortunate enough to be able to afford a telephone, by which she may communicate with the tradespeople at any minute, realizes what a blessing this servant of electricity has become to mankind. In cases of illness, in troubles of any kind, aid and friends are so easily summoned that fear seems almost a thing of the past. What telephone communication with the outside world is to the house, so is the Inter-phone system to the members of the household and to the proper, economical and comfortable running of a home.

—Book of Home Building and Decoration, *1912*

fellows come along and seek to complicate matters."

Those who did want and get one acquired a clunky wooden wall instrument. (Congressman James A. Garfield, later to be president, was one of the earliest subscribers.) But in short order, the telephone became a more mobile device. Attitudes toward it varied. In some homes, it was accorded a place of honor: a pedestal in the parlor or its own table in the hall. In other homes, it was an offensive interloper to be hidden away in a cupboard or beneath a handmade "telephone doll," cousin to a tea cozy. Whatever the instrument's social status in the

Three coeds share the line in MGM's Spring Madness, *1938.*

other agreed with Father—she didn't like telephones either. She distrusted machines of all kinds; they weren't human, they popped or exploded and made her nervous. She never knew what they might do to her. And the telephone seemed to her, and many other people, especially dangerous. They were afraid that if they stood near one in a thunderstorm they might get hit by lightning. Even if there wasn't any storm, the electric wiring might give them a shock. When they saw a telephone in some hotel or office, they stood away from it or picked it up gingerly. It was a freak way to use electricity, and Mother wouldn't even touch the queer toy. Besides, she said, she had to see the face of any person she talked to. She didn't want to be answered by a voice coming out of a box on the wall.

—*Clarence Day,*
Life With Father, 1948.

household, everyone realized that a miracle—as yet imperfect—was at hand. The act of telephony was still cacophony to the ears. Static, sputtering, clatter accompanied every call. But none of that mattered when the wavering voice of Aunt Fanny could be heard from ten miles away.

When the telephone rang, friends and family gathered round, as mesmerized by its magic flow of

Another hall abomination is a telephone. Unless we want our guests to know the price of their roast, or the family to listen aghast while we tell a white lie for society's sake, or the cook to hear us asking for a new one's references, don't put your telephone in the hall. Closet it, or keep it upstairs, where the family alone are the bored "listeners in."

—*Agnes Foster Wright,* Interior Decoration for Modern Needs, *1917*

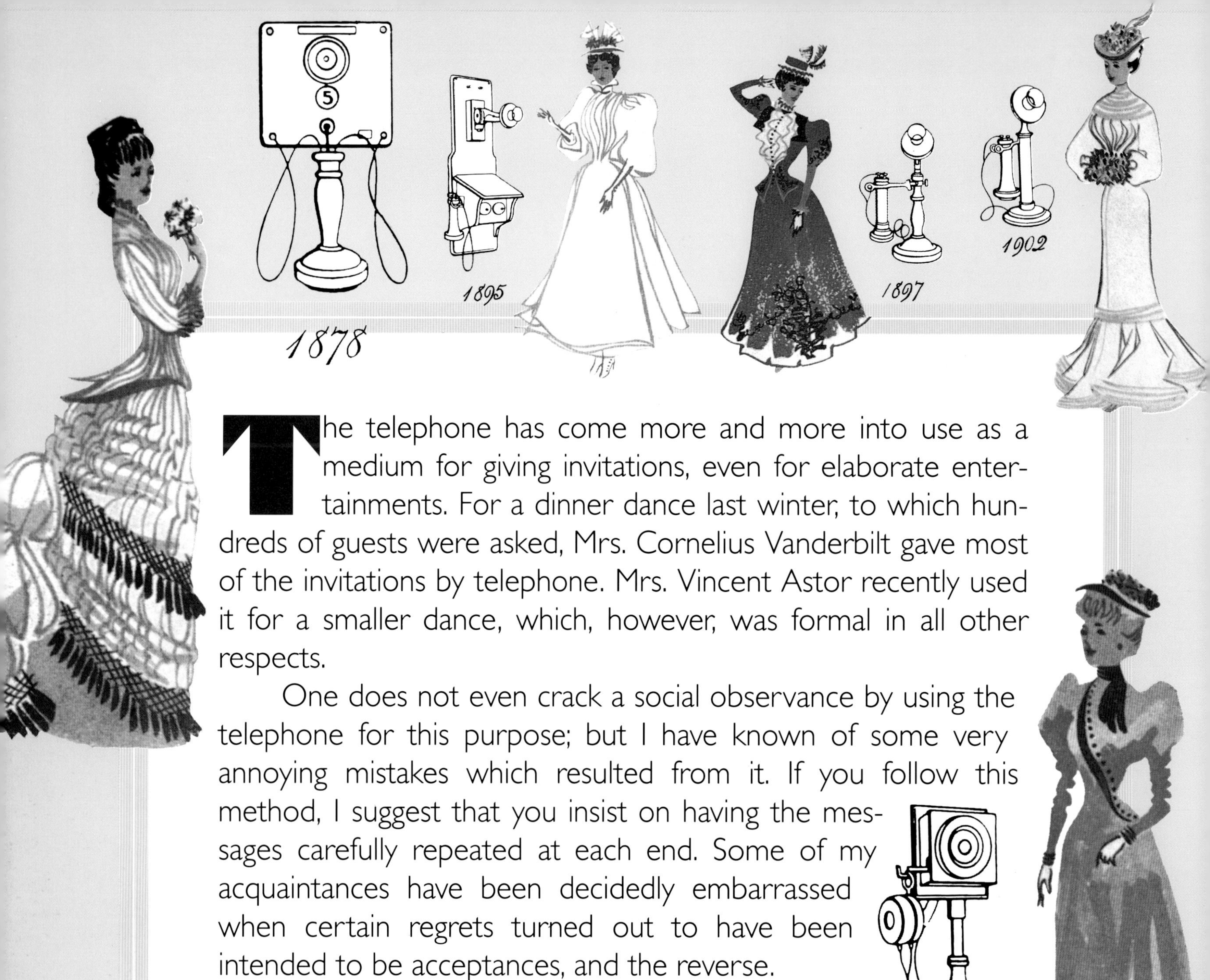

The telephone has come more and more into use as a medium for giving invitations, even for elaborate entertainments. For a dinner dance last winter, to which hundreds of guests were asked, Mrs. Cornelius Vanderbilt gave most of the invitations by telephone. Mrs. Vincent Astor recently used it for a smaller dance, which, however, was formal in all other respects.

One does not even crack a social observance by using the telephone for this purpose; but I have known of some very annoying mistakes which resulted from it. If you follow this method, I suggest that you insist on having the messages carefully repeated at each end. Some of my acquaintances have been decidedly embarrassed when certain regrets turned out to have been intended to be acceptances, and the reverse.

—*Mrs. Lydig Hoyt, "How to Avoid Social Blunders,"*
The American Magazine, *January 1922*

electrons as they would later be by the radio. Making a call also drew a crowd, but use was permitted only if the actual caller was a registered subscriber. "Annie Kate! What are you doing on the McGillicuddys' line?" the central exchange might admonish a chatterbox using her neighbor's phone. "You know the rules. If you don't get off this line, I'll have to report it to the Company, and they'll come out and take that telephone away." The operator recognized the voices of her customers. She also was a stickler about early Company policy. If it wasn't your phone, you couldn't use it.

As more and more people wanted more and more phones, even if they couldn't afford one of their very own—

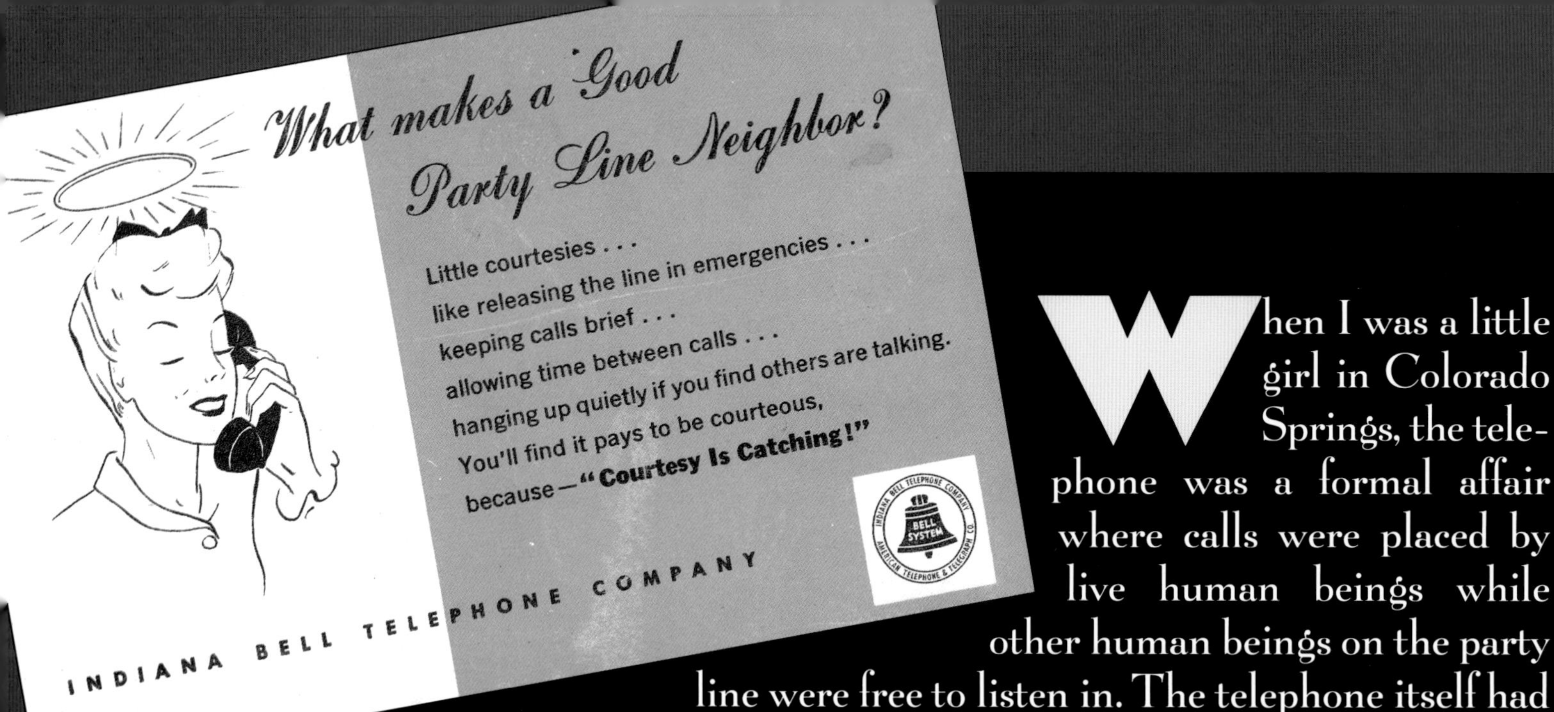

When I was a little girl in Colorado Springs, the telephone was a formal affair where calls were placed by live human beings while other human beings on the party line were free to listen in. The telephone itself had the stature of a Shinto shrine in our house. As in most forties houses, the Tokonoma corner for the telephone was built in, giving the home owner no choice for its placement.

The telephone had its own kind of anticipatory power. Solid black, slightly warm to the touch like amber, and nice to hold like a well-balanced garden tool, it had four different rings for each party-line member. Two short blasts for us, one for the Halleys down the street, and so forth. My mother loved to guess who was calling the other lines, and I suspect that she listened in now and then to see if she was right. We, the children, were subject to strict rules about never answering any but our ring. Because of the party line, the phone was not used to chat for hours or to reveal private matters. When someone on the line picked up the phone, it was a signal to cut the conversation short so others could use it. The telephone had the power of ritual in those days, before it became private and dialed and ubiquitous.

—Carolyn Reynolds Ortiz, television producer, 1993

and as telephone wires were being strung hither and yon—families joined the party line. In rural areas, there was no other choice. Two to twelve parties might share a single number, each with its own identifiable ring, which could be heard in every participating household.

But all parties didn't share a sense of privacy. In spite of newly minted telephone etiquette, monopolizing the line and listening-in became popular and irresistible pastimes. Recalls one Kansas lady, "People quickly learned everyone's ring, and if one was a particularly colorful or controversial individual, it was tempting to 'pike in on' the conversation. There were few secrets among people on a party line."

Early in the century, telephony was all the rage. Ephemera of the time reflected its popularity. It was

"ARE YOU THERE"

"We're having some people," everybody said to everybody else, "and we want you to join us," and they said, "We'll telephone."

All over New York people telephoned. They telephoned from one hotel to another to people on other parties that they couldn't get there—that they were engaged. It was always tea-time or late at night.

—*Zelda Fitzgerald,*
Save Me the Waltz, *1953*

The Farmer and the 'Phone

To the farmer, the telephone means more than any of the great inventions of the nineteenth century, especially when he can build his own line, as in this case the expense is practically nothing compared with the benefit.

Some Reasons Why

☎ It saves time, "horseflesh," and money, making many trips to the village unnecessary.

☎ It gives the farmer all the advantages of his village neighbor, by placing him in communication and easy access of all.

☎ It orders supplies from his hardware or implement dealer sent out by parties coming in his direction, and in urgent cases by special messenger, saving the time which to the farmer in seedtime and harvest means many dollars.

☎ It gives him the daily weather reports whenever he cares to inquire, enabling him to avoid loss of crops by storm, and the opportunity of planning his work accordingly.

☎ It enables him to take every advantage of the market in the sale of the grain, and his grain buyer is ever ready to keep him informed. In this alone, he can save more than the cost of his telephone every year.

☎ It enables him to call up his grocer, and sell his butter, eggs, and vegetables before they leave the farm, receiving therefore an average price far in excess of what he would receive were he compelled to accept what is offered in a congested market.

☎ It oversteps storms and snowdrifts, and brings to him assistance in time of need.

☎ It renders fire protection, and is the best "thief catcher" in the world.

☎ It is protection to wife and daughters against the importunities of tramps and vagabonds, and gives a security nothing else can.

☎ It finds the strayed cattle, returns the lambkin to the fold, and becomes the shepherd of the neighborhood.

—Telephone Hand-Book, *compiled by H. R. Van Deventer, 1910*

A Telephone On Your Farm

will actually save you money—you can phone to town for prices and sell at the highest figure—find out about the weather—order supplies, etc. A telephone will make life pleasanter for your wife and children, too. Talk to neighbors any time.

It's easy for any group of farmers to get together and build their own line.

Let us explain how—send for the Free Bulletin:

"A Telephone on the Farm"

It tells how very cheap it is to build and run your neighborhood system, how to organize a company; gives by-laws and rules. Write today.

STROMBERG - CARLSON TEL. MFG. CO.

Independent Telephones. **Address Dept. A, Nearest Office, Rochester, N. Y., Chicago, Ill., Toronto, Can.**

The subscriber has a right to expect the first word from the operator to be always "Number?" to which the word "please" had better be added, but is not absolutely required.

The subscriber has the right to expect the operator, if necessary, to say, "That line is busy"; simply "Busy" won't do.

The operator has a right to expect that the subscriber will have the number ready when the operator answers, and that the operator will not be compelled to wait until the subscriber looks it up in the directory.

Also that the subscriber will give the number in a clear and distinct voice, and if the operator misunderstands a number, that she will be corrected, without evidence of anger in the tone of the subscriber.

—Telephone Etiquette, *1905*

Points to Remember

Treat the operator politely.

Be polite in speaking to friend and stranger alike.

If a wrong number is given, do not be impatient or rude.

If you are called to the phone, reply quietly.

If you have called the wrong number by mistake, apologize for the trouble you have given.

Be brief.

The one who has called up is the one to close the conversation.

Do not talk about private affairs.

Never "listen in." It is dishonorable.

Do not try to get the line away from another who is using it except in extreme cases. Apologize if you feel compelled to ask for the line.

Do not say things you would not say in talking face to face.

—*Nancy Dunlea,* The Courtesy Book, *1927*

The Little Matter of Replacing the Receiver

When you have finished your telephone visit, and courteously said "Good-by" or "Thank you," replace the receiver gently. Slamming the receiver might cause a sharp crack in the ear of the person with whom you have been talking. Since you would not "slam the door" after an actual visit, be just as careful in closing your telephone door.

—*The New York Telephone Company,*
You and Your Telephone, *1940s*

a new age in graphic design. In magazine advertising and on wildly popular picture postcards, the telephone was being featured as prominently as the cigarette and automobile would be thirty years later. Tin Pan Alley produced hundreds of telephone tunes such as "A Ring on the Finger Is Worth Two on the Phone," "Call me Up Some Rainy Afternoon," "Hello, Central, Give Me Heaven," and "My Lady of the Telephone."

Lady of the telephone indeed. The phone and the American woman were made for each other.

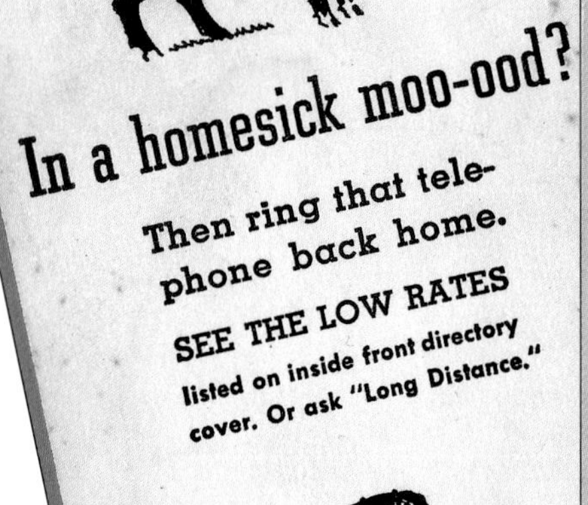

IN
STORMY
WEATHER
USE THE
TELEPHONE

The Telephone as an Instrument of Self-Torture

The telephone, when you do speak on it, effectively strips your personality of all its non-audio charm—all smiles and winks and other facial expressions that help to convey subtlety and clarify your meaning. And God help you if you don't have a beautiful voice.

Then, too, when you speak on the telephone you never know quite what's going on at the other end of the line. You can't see the facial expressions of whomever you're talking to, so you really never know where you are with them. Perhaps they're bored. Or dripping wet. Or watching television. Perhaps they've put the receiver down and walked away. Perhaps someone else is with them, listening to what you're saying, and they're both exchanging funny faces and other signals, and they can scarcely contain their laughter about what you're saying.

—*Dan Greenburg and Marcia Jacobs,* How to Make Yourself Miserable, *1966*

Free Hand Syndrome (FHS), *ailment affecting the neurovascular bundle in the brachial-plexus region of people who talk on the telephone while simultaneously opening mail, cooking, taking notes, etc.*

According to the doctor who coined the term, Jacob D. Rozbruch, chief of orthopedic surgery at New York's Beth Israel Medical Center North, anyone who habitually cradles the receiver between ear and shoulder in a sort of one-sided shrug is at risk for developing FHS. And no wonder. "The position is an extremely awkward one for the neck," says Dr. Rozbruch. "You can't expect the body to accept such an arrangement peacefully." As he explains it, twisting the head and neck like this stretches the ligaments of the cervical spine. Help! cry the ligaments. Quick to the rescue, the nearest nerve picks up the SOS and communicates it to the brain which orders the appropriate muscles to compress in order to prevent further trouble. The result: tightness and spasm which, if ignored, can lead to nerve damage and chronic pain that sometimes only surgery can relieve. "The only real remedy is to stop hurting yourself and figure out a better way," advises Dr. Rozbruch. "Get a headset. The regular telephone was simply not designed to use without a hand."

—*Kathryn Rose Gertz, writer, 1993*

Gregory Peck as Captain Newman, M.D. (Universal, 1965).

"I'll call you up!"

Isolated farm wives and housebound homemakers were no longer limited by the length of their apron strings. The telephone line extended their boundaries—to the grocery store in stormy weather, the doctor and fire department in times of distress, Father's office for his estimated time of arrival, Miss Rose's Tea Room for luncheon reservations, and, of course, to gossipy Gertrude eager to chew the fat and spill the beans—for hours and hours and hours. As Ring Lardner observed at the Algonquin round table in 1923, wives were "people that think when the telephone bell rings it is against the law not to answer it."

Walter likes to watch games on television—basketball, baseball, the horse races—anything he happens to be betting on. And now with cable, one can watch games twenty-four hours a day. Dinner is a time I usually like to talk about everything and nothing, big things and little things, but now I notice that if he thinks I have something to say he instantly turns the television on. At first I tried to do a few little things that would hint to him how rude I think that is, but he didn't pay much attention. Then I tried more things; I even gave myself a pedicure during dinner one night. He didn't seem to notice.

So now I simply leave the table and go upstairs and have dinner in bed and telephone anyone I know who is in any foreign country and speak to them for hours. He doesn't seem to notice this either.

He'll say a little something or two like, "You're always on the telephone."

To which I'll reply, "You're always on the television." And that is that.

—*Carol Matthau,* Among the Porcupines, *1992*

CUPID'S MESSAGE
Sweetest story ever told.

Chapter 6

Love on the Line

Before the phone call, there were bouquets and billets-doux, singing telegrams and moonlight serenades. But these tokens of affection were often complicated to arrange and not always gracefully accepted. The telephone provided a safety net, offering intimacy and invisibility at the same time. Once its static cleared, shy Henry could whisper sweet nothings into Elsie's ear without having to look her in the eye. George didn't have to shave his whiskers before calling Dora, and Dora didn't have to put up her hair to take his call. And the working-class heroine of a 1927 Broadway musical, *The Five O'Clock Girl,* could snag her polo-playing prey over the wires by assuming an upper crust accent.

"The telephone is really the life-line of romance," says the eternal *Cosmo* girl, Helen Gurley Brown. But it is also as taut as a leash, binding the lovesick to its bidding. Will it ring? Won't it? Why doesn't it? In her anguished short story, "The Telephone Call," Dorothy Parker summed up the torture familiar to all

(Top) *Jayne Mansfield holds on.*

women who ever waited by the phone . . . waited for Him to call. And if it does ring, will it bring a smile or break a heart? The tear-drenched telephone is far more than a prop for Cocteau's lady in *The Human Voice* and Luise Rainer in *The Great Ziegfeld.*

The phone has always been cupid's agent, the means of facilitating an innocent date for tea or an illicit one for love in the afternoon. It is the medium

The telephone is handy for a novelist, since it forces his characters to fill silences with stories and so does some of his work for him. But *Vox* is, despite its safely sexual "adult theme," a product of my own childhood infatuation with the telephone. I was interested back then in the sensations of dialing, of having my finger guided in its numerical hole (first it was black metal, then more comfortable clear plastic) along arcs of a perfect circle, as if it were a pen in a Spirograph.

Also, for a period of several years while I was growing up, no member of my family wore a watch, and our house had no working clock. My job was to call, often several times a day, the time and temperature number (a service then provided by Rochester Savings Bank) and find out what time it was. I was proud to make these calls. The other phone numbers I had memorized merely reached people my own age (e.g., my friend Fred, GI 2-1397, and my friend Maitland, CH 4-4158); but the time-temperature number linked me to a more real, kitchenless world of Cesium clocks and compound interest and absolute zero, to times and temperatures thrillingly beyond dispute, endorsed, it seemed, by the National Bureau of Standards and the FDIC. The day after daylight savings, the time and temperature number was always busy, a sign of simultaneous citywide activity as definite as the drop in water pressure during the ad-breaks in the Super Bowl.

A final relevant phone-happiness came in learning the trick of calling myself: you dialed some short number and you made a carefully-timed click of the cradle, and miraculously, your own phone, the phone you were touching, would ring, a result that seemed, in those years before the discovery of other solitary auto-dialed pleasures, exotic and shocking and worthwhile.

It isn't stretching things too much to say that in *Vox* I was performing the novelistic equivalent of these early telephonic diversions: I was calling up, or calling on, what I hoped were National-Bureau-of-Standards-level verities about the interests and flirtations of two representative single phoners, a pair who began as strangers to me and to themselves and who thus had to move as mere voices from the absolute zero of their initial connection to the high Fahrenheit range of their affectionate spoken orgasms. And at the same time, of course, I was making my own phone ring.

—Nicholson Baker, 1993

"Whoo! How long have we been talking?"

"Hours and hours."

"Hours and hours and hours," she said. "My mouth is chapped. Too much making out."

"Is your voice sore?"

"It really is. Whoo! Gee, I'm going to have to call in sick *again*. I'll sleep all day, mm, sounds delightful. The hiss on the phone is very loud now, isn't it? That companionable hiss. It's always louder at the end of conversations."

"Oh, is it the end already?" he said. "Couldn't we just fade out somehow, talking and talking? I can't think of a better way to invest my life savings. Not that I'm much of a saver."

"You're quite a telephoner, though."

"You are too! I mean it! I think really this is one of the nicest conversations I've ever had."

"I liked it too," she said. "I don't know, though—do you think we talked enough about sex?"

"Not nearly enough. I—"

"Yes?" she said.

"Do you think our . . . wires will cross again?"

"I don't know. I don't know. What do you think?"

"I could give you my number," he said. "I mean if you still want it. I'll avoid a possibly awkward moment by not asking for yours. Or we could meet out here again, if you'd rather do that."

"Out here under the stars? I can't afford it. Where's a pencil? Ah, a nice blunt pencil. Tell me your number."

He told her. She read it back to him.

"Call me soon," he said. "In fact, call me in a few hours, after you've topped yourself off in the shower."

"You know me too well."

"I like you a lot."

"I wonder what you look like," she said.

"Surprisingly normal. Maybe someday you'll know."

"It's a possibility."

—*Nicholson Baker, Vox, 1992*

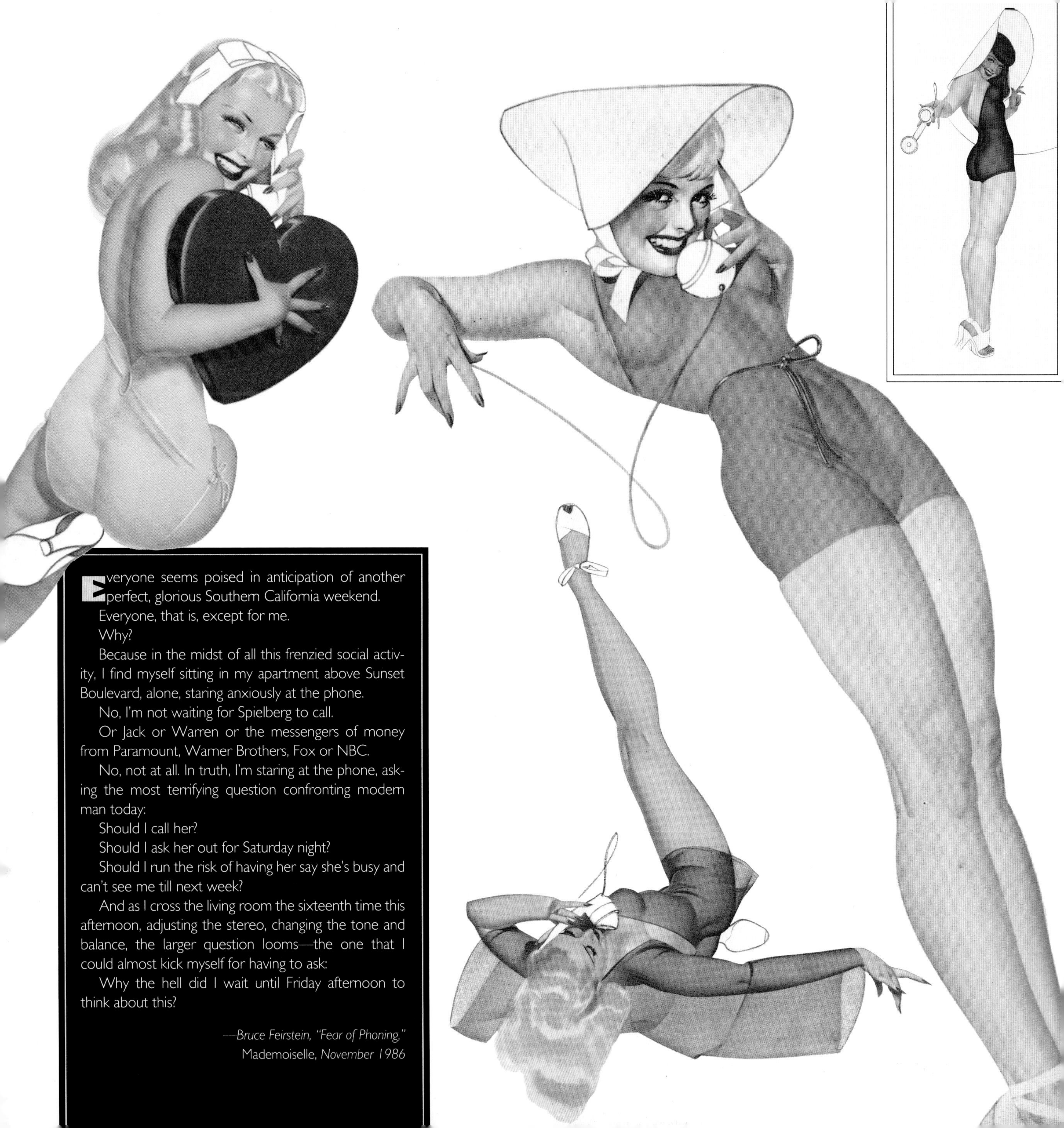

Everyone seems poised in anticipation of another perfect, glorious Southern California weekend.

Everyone, that is, except for me.

Why?

Because in the midst of all this frenzied social activity, I find myself sitting in my apartment above Sunset Boulevard, alone, staring anxiously at the phone.

No, I'm not waiting for Spielberg to call.

Or Jack or Warren or the messengers of money from Paramount, Warner Brothers, Fox or NBC.

No, not at all. In truth, I'm staring at the phone, asking the most terrifying question confronting modern man today:

Should I call her?

Should I ask her out for Saturday night?

Should I run the risk of having her say she's busy and can't see me till next week?

And as I cross the living room the sixteenth time this afternoon, adjusting the stereo, changing the tone and balance, the larger question looms—the one that I could almost kick myself for having to ask:

Why the hell did I wait until Friday afternoon to think about this?

—*Bruce Feirstein, "Fear of Phoning,"*
Mademoiselle, *November 1986*

and the massage, without which the call girl could not ply her trade, the 800 or 900 number play its multi-million-dollar tune, nor Nicholson Baker write *Vox.* His erotic best-seller, about tender but electronic sex between two strangers patched in randomly by a central switchboard, takes place entirely on the phone.

Intimacy and invisibility then, intimacy and invisibility now.

Hello, Flo? Yes, it is Anna. I am so happy for you today. I could not help but calling you and congratulating you. Wonderful, Flo. Never better in my whole life. I am so excited about my new plans. I am going to Paris. Yes, for a few weeks, and then I can get back and then I am doing a musical, and I—oh, it's all so wonderful. And I'm so happy. Yes, and I hope you are happy, too. Yes. Oh, I am so glad for you, Flo. It sounds funny for exes, but every ex-wife could tell each other how happy they are, oui? Yes, Flo. Good-bye, Flo. Good-bye, my darling.

—Luise Rainer as Anna Held, smiling through her tears as she congratulates Florenz Ziegfeld, her former husband, upon his marriage to Billie Burke.
The Great Ziegfeld, *1936*

Love on the Line

If you lead a fast, frantic Manhattan life, most likely whatever romantic flames you have blazing are to a certain degree, from necessity, kept fanned through the telephone.

But what is the appropriate etiquette of telephone romance?

As Charlotte Ford sees it, she of the recent book on modern manners, in today's business world "probably several calls during the day, rather than one long call, would be preferable."

"Two or three minute breaks in the day," she decides from her dress-designing office on Seventh Ave. "I think it's better short and sweet."

The real zinger, it is generally agreed, is the wildly appreciative laugh. It shows such an interest. So necessary when operating without eye contact. "It's the telephone equivalent of the seductive smile," explains one telephone jockey. At least it is if you have a charming laugh."

"Even better," says this authority in her chosen field, "is really giggling. The telephone is the perfect medium for telling jokes, which are aural, and for just staying light and gay—I mean, when you think of the two of you holding black instruments in your hands and talking animatedly into them, with heart—well, there is a special silliness to it."

—Jan Hodenfield, New York Daily News, *January 29, 1981*

A Chat with Helen Gurley Brown

Q: Do women still wait by the phone?

HGB: Well, they try *not* to. They all have answering machines, so it's not *necessary* to wait by the phone, but I think psychologically you could say that they are still waiting by the phone, because even if they go out to get a pack of chewing gum they're hoping that when they get back there'll be a message.

Q: What about boys calling girls and girls calling boys?

HGB: I don't think anything has changed dramatically since 1912. You can call a man to ask him something, or to thank him for something—to do sort of *business*-like things—but I think if the call is to be a romantic one, it's got to be him calling you. Just to call up and say, "I was thinking about you"—you can do that with a husband, or you can do it with somebody that you like better than he likes you—we always know who's who. So if it's a man who just will keel *over* when he hears from you, okay, be my guest, you can be indulgent and call him up if you want to, just to give him a real . . . jolt, a real happy moment. But that's for *that* kind of person. That's not for the man that *you're* crazy about. You don't call *him* and say, "Oh, I was just thinking about you, just wanted to hear your voice." A little of that goes not only a *long* way, it goes all the way around the world.

It turned out that the reason a young woman named Helen Gurley (who later married me) didn't return my calls was that she put her telephone in the refrigerator when she went to sleep—and often forgot to take it out.

—David Brown,
film producer, 1993

What to Call a Call Girl

Whenever a new girl joined up, the first thing she needed was a working name. Rather than assign one arbitrarily, I would let her choose her own favorite, and if the name she selected was at all suited to her looks and personality, she was welcome to use it. While I tried to be flexible, I had to draw the line when a bouncy, blond blue-eyed art student insisted she'd like to be known as Natasha.

To me, certain names have always suggested specific images. A girl named Natalie should have long dark hair. Alexandra is tall and stately, and Ginger is a spunky redhead. I gave the name Melody to a girl named Carol whose voice was so mellifluous it made me think of music. When Wilma joined up, tall, sophisticated, and experienced, I knew immediately that she would be Claudette. I encouraged foreign girls to choose familiar names from their country of origin: Sonya from Germany, Kristen from Scandinavia, Gabriella from Brazil.

These new names weren't only for the comfort and convenience of the girls; they were equally important to those of us who had to describe the escorts over the phone. When I or one of the assistants was describing a girl to a client, we would have to conjure up an image of her in just two or three phrases. Her name could be a big help in that process.

—*Sydney Biddle Barrows,* Mayflower Madame, *1986*

Phone sex is phony sex. That's the joy of it. You get to be whoever you want to be, which is the way sex should always be (our made-up selves being our truest selves).

I worked for a fantasy switchboard over a few weeks one winter because I was writing a thriller with phone sex as the theme. *Hotline* is stalled on page 50, but never mind. We took the slump out of February, my phonefux and I.

On a typical evening, the central office (which I never saw) would arrange for me to make four or five calls. I might be a bosomy tart named Ginger, fifteen-year-old Alicia with white cotton panties and a shy little voice, relentless Mistress Dominita, sincerely solicitous Sondra, or whatever other self the situation required. The "Larry" or "Rod" or "Roberta" at the other end got twenty minutes of my torpid attention, for which precisely calibrated pleasure he would pay the switchboard fifty dollars, with a fraction going to me.

Big claims are made for phone sex (that it keeps people from getting AIDS) and against (that it degrades). Maybe both sides are right—and that's part of the kick. To save lives and be degraded in one is no small delight.

The romp ended when Roberta, a transvestite from Boston, sent me pictures of herself (via the switchboard and the P.O. box I'd rented in the name of Vicki White, the *Hotline* heroine). Those pictures broke the connection. Costumes of the body are nothing next to costumes of the mind. You see so much more mouth-to-ear than eye-to-eye. I had loved Roberta the true; I pitied Roberta the real. Sex-for-pity is a treacherous thing. I sighed and hung up.

—Nancy Weber, novelist, 1993

You remember the telephone when you were courting me? I remember everything when I was courting you even the gander that used to rush and hiss at me when I took you in my arms. You remember the telephone when you were courting me silly? I remember. Then you remember the party line going eighteen miles along Cole Creek Valley and only five customers? I remember I remember the way you looked with your big eyes and your smooth forehead you haven't changed. You remember the telephone line and how new it was? Oh it was lonely out there with nobody in three or four miles and nobody really in the world but you. And me waiting for the telephone to ring. It rang two times for us remember? Two rings and you were calling from the grocery store when the store was closed. And the receivers all along the line all five of them going click-click Bill is calling Macia click-click-click. And then your voice how funny it was to hear your voice the first time over a telephone how wonderful it always was.

"Hello Macia."

"Hello Bill how are you?"

"I'm fine are you through with the work?"

"We just finished the dishes."

"I suppose everybody is listening again tonight."

"I suppose."

"Don't they know I love you? You'd think that was enough for them."

"Maybe it isn't."

"Macia why don't you play a piece on the piano?"

"All right Bill. Which one?"

"Whatever one you like I like them all."

"All right Bill. Wait till I fix the receiver."

And then way out on Cole Creek way west on the other side of the mountains from Denver music tinkling over telephone wires that were brand new and wonderful. His mother before she was his mother before she thought particularly of becoming his mother would go over to the piano the only one on Cole Creek and play the Beautiful Blue Ohio or perhaps My Pretty Red Wing. She would play it clear through and his father in Shale City would be listening and thinking isn't it wonderful I can sit here eight miles away and hold a little piece of black business to my ear and hear far off the music of Macia my beautiful my Macia.

"Could you hear it Bill?"

"Yes. It was lovely."

Then somebody else maybe six miles up or down the line would break into the conversation without being ashamed at all.

"Macia I just picked up the hook and heard you playing. Why don't you play After the Ball is Over? Clem'd like to hear it if you don't mind."

His mother would go back to the piano and play After the Ball is Over and Clem somewhere would be listening to music for maybe the first time in three or four months. Farmers' wives would be sitting with their work done and receivers to their ears listening too and getting dreamy and thinking about things their husbands wouldn't suspect. And so it went with everybody up and down the lonesome bed of Cole Creek asking his mother to play a favorite piece and his father listening from Shale City and liking it but perhaps growing a little impatient occasionally and saying to himself I wish the people out on Cole Creek would understand that this is a courtship not a concert.

—*Dalton Trumbo,* Johnny Got His Gun, *1939*

Music to My Ears

Considering that the telephone only burst into the world's incredulous consciousness at the great Philadelphia Centennial Exposition of 1876, its arrival in the musical world was remarkably swift. Those glorious British madcaps William S. Gilbert and Arthur Sullivan (neither of them had yet arrived at "Sir"dom) unleashed their *H.M.S. Pinafore* in 1878. At the climax of this great spoof of mistaken identities and political influence-peddling, the hero Ralph must be sent into prison in the ship's brig. A dismal and isolated place it must, indeed, be; as the villainous dick Deadeye tells us:

"He'll hear no tone/Of the maiden he loves so well!
No telephone/Communicates with his cell!"

As our hero might remark, "Horror!"

Half a century later we come across yet another benighted locale, made all the bleaker by its absence of telephone. At Germany's avant-garde Baden-Baden music festival in the summer of 1927 the great hit was a surrealistic cantata by another proficient words-and-music team. In Bertolt Brecht's words, the denizens of the fictitious state known as Mahagonny have grown bored with their Utopian existence. Worst of all, their outcry, to the tragicomic music of Kurt Weill, includes the verbal exchange

"Is there no telephone?"
"Is there no telephone?"
"No!"
"O, sir, God help me!"

But the stage-piece that truly brought operatic heroism to the telephone came two decades later, in the two-character opera-in-brief by Gian-Carlo Menotti called, simply enough, *The Telephone*. It opened on Broadway in 1947, as a curtain-raiser to the more serious *The Medium*, enjoyed a respectable run, and lives on as a charmer within the reach of amateur music societies.

Ben and Lucy are in love, but you'd hardly know it. Lucy spends her life on the telephone, with Ben wringing his hands in the corner, trying in vain to get a word in edgewise. "Hello, hello," she trills in her one big aria, "Oh Margaret, It's you!" and on and on. Finally Ben comes up with a solution. Down to the corner phone booth he goes, puts in his nickel (this is 1947, remember), and dials his sweetie. She answers, he proposes, happy ending.

If you can find the out-of-print LP of *The Telephone*, or if it ever turns up on CD, that "Hello, hello" aria makes a neat answering machine message. I know; I used it for a while, and as a reward one day there was a message from none other than Beverly Sills. "Hello, hello," she trilled, "it's BEV-er-ly." You can have your Callas and Caruso records; my treasure is Beverly on the telephone.

—Alan Rich, music critic, 1993

Please, God, let him telephone me now. Dear God, let him call me now. I won't ask anything else of You, truly I won't. It isn't very much to ask. It would be so little to You, God, such a little, little thing. Only let him telephone now. Please, God. Please, please, please.

If I didn't think about it, maybe the telephone might ring. Sometimes it does that. If I could think of something else. If I could think of something else. Maybe if I counted five hundred by fives, it might ring by that time. I'll count slowly. I won't cheat. And if it rings when I get to three hundred, I won't sotop; I won't answer it until I get to five hundred. Five, ten, fifteen, twenty, twenty-five, thirty, thirty-five, forty, forty-five, fifty . . . Oh, please ring. Please.

This is the last time I'll look at the clock. I will not look at it again. It's ten minutes past seven. He said he would telephone at five o'clock. "I'll call you at five, darling." I think that's where he said "darling." I'm almost sure he said it there. I know he called me "darling" twice, and the other time was when he said good-by. "Good-by, darling." He was busy, and he can't say much in the office, but he called me "darling" twice. He couldn't have minded my calling him up. I know you shouldn't keep telephoning them—I know they don't like that. When you do that, they know you are thinking about them and wanting them, and that makes them hate you. But I hadn't talked to him in three days—not in three days. And all I did was ask him how he was; it was just the way anybody might have called him up. He couldn't have minded that. He couldn't have thought I was bothering him. "No, of course you're not," he said. And he said he'd telephone me. He didn't have to say that. I didn't ask him to, truly I didn't. I'm sure I didn't. I don't think he would say he'd telephone me, and then just never do it. Please don't let him do that, God. Please don't. . . .

Why can't that telephone ring? Why can't it, why can't it? Couldn't you ring? Ah, please, couldn't you? You damned, ugly, shiny thing. It would hurt you to ring, wouldn't it? Oh, that would hurt you. Damn you, I'll pull your filthy roots out of the wall, I'll smash your smug black face in little bits. Damn you to hell.

—Dorothy Parker, "A Telephone Call," 1930

WAITIN' FOR A CALL FROM YOU

When a man calls you from Tulsa, he invariably makes the mistake of calling either from a public bar or from his mother's living room. Neither setting is exactly conducive to a free exchange of ideas. There, within earshot of his fellow revelers or his mother, he can hardly say the one thing you want to hear, which is that he misses you terribly, it's been a nightmare, a nightmare! and he's never going to make a trip alone again. For that matter, you can't tell him you miss him either, because the children are there with you and they become downright alarmed at any hint that their parents have preserved this degrading adolescent attachment so far into senility. So, if you're not careful, it's going to be a total loss of five dollars and eighty-five cents.

Don't, whatever you do, launch into that foolish litany of last-minute health bulletins: "Yes, I'm fine, yes, Chris is fine, yes, Gilbert is fine, etc." Let it be understood in advance that if one of the children should be rushed to the hospital for an emergency appendectomy, you'll mention it.

Use the time to clear up some matter that has really been troubling you. Explain that you finally saw *The Bridge on the River Kwai* on television and that it was marvelous, marvelous, but you didn't understand the ending. Get him to explain it. Did Alec Guinness mean to set off that dynamite or didn't he? What about William Holden? Who really killed him? This is important. When William Holden gets shot, a woman wants to know the facts. Later, when you hang up, you may discover that you've forgotten to ask what time his plane arrives at La Guardia, but the call won't have been a total loss.

—*Jean Kerr,* How I Got To Be Perfect, *1958*

(Top, left) *Bette Davis.* (Above) *A Valentine's Day window display at Tiffany's, 1984.* (Left) *Bogie and his baby in Warner Bros.'* The Big Sleep *(1946).*

Do you know who I've always depended on? Not strangers, not friends. The telephone! That's my best friend.

—*Marilyn Monroe*

It has to be BIG!

Chapter 7

"Don't Call Us, We'll Call You."

"There's something about saying 'OK' and hanging up the receiver with a bang that kids a man into feeling that he has just pulled off a big deal, even if he has only called up Central to find out the correct time," Robert Benchley observed one day to his cronies at the Algonquin Round Table. They all nodded. They all knew.

Phone power. It can be used or abused, depending on who's making or taking the call. It's *The Front Page,* with a wire directly to the Criminal Courts Building—or it's Woodward and Bernstein, with a line on the Oval Office. It's the telemarketer, hawking a swampy acre in the Everglades, or the computerized voice putting you on permanent hold. It's the delinquent suitor who hangs you up or the cranky plumber

(Above, right) *Charles Bronson brandishes his* Telefon *in MGM's 1977 thriller.*
(Right) *Fredric March and players in Columbia's* Bedtime Story *(1941).*

"What? My wife has given birth to a baby boy? Good, put him on the phone!"

Telephones are not necessarily a convenience—for many they are a prop. I know a good many people who arrange to have their secretaries call them at lunch so that a telephone can be brought to the table and plugged in, and what's more, won't eat in a restaurant where they can't have a phone at the table. One top executive in New York has a private bathroom with a telephone on the wall next to the toilet, which is certainly a convenience for a busy man, but not a very public symbol of power. Another major corporation executive has telephones concealed in small green, rustic boxes attached to trees on his estate, so that he can make calls and receive them even when he's walking to the pool or the boathouse, the insistent ringing rising above the gentle woodland noises of the birds and the wind. Once people have associated telephones with prestige, there's no end to what they can do. There are radiotelephones built into handsome leather brief cases, which sound a soft, distinctive hum when you're being called, and sell for just over $2,000—well worth it in case somebody rings you on your way from your office to your "limo." More plebeian telephone addicts carry rolls of dimes, fresh from the bank, in their pockets, and set up shop in telephone booths, desperately anxious "to keep in touch" at all times.

The telephone is a perfect example of how we make do with what we've got to create power symbols. What was invented as a mundane and unattractive convenience, we have made into a complex mark of status and power, as if by instinct. If we have a visitor in our office, we can demean him by accepting telephone calls while he's talking, or impress him by saying, "Excuse me, it's the chairman of the board calling"—or the President of the United States, or "the Coast" or a call from overseas—and finally, if we want to flatter him, we can say, "Hold *all* my calls, whoever it is." Nothing puts a person in their place better than carrying on a dialogue with a man who has a telephone receiver cocked between his ear and his shoulder, and who says, "Keep right on talking, I'm listening, I just have to take this call."

—*Michael Korda,* Power!, *1975*

(Right) Paul Muni as the Counselor-at-Law (Universal, 1933)

who hangs up on you. It's the kidnapper who calls at a designated hour or the agent who won't take your call at any hour. It's the sleuth and the pollster, the bigwig's secretary and the extortionist, the unseen villain terrorizing Barbara Stanwyck. It's the enraged business partner launching a verbal attack without fear of face-to-face confrontation and the obscene caller hiding behind a cloak of vile invisibility.

The ubiquitous device that serves the gossip, emboldens the lover, and sustains the teenager also has a dark side. As Raymond Chandler's cynical gumshoe, Marlowe, sums it up: "There is something compulsive about a telephone. The gadget-ridden man of our age loves it, loathes it, and is afraid of it."

The phone rings . . . and it must be answered.

(Right) *Ted Wass in MGM/UA's* Curse of the Pink Panther *(1983).*

The Tyranny of the Telephone

The New York Stock Exchange did business for nearly a full century before the telephone was invented; and you wonder how they built the railroads, stretched the country across a continent, got married, and raised families without the telephone. But they did. In fact, Shakespeare wrote *Hamlet,* and Mozart even composed *Don Giovanni* without the help of the phone.

There's something about it that only a trained psychologist could explain. You receive a letter and you either open it or leave it unopened, as you wish. You put it in your pocket, or in your apron, or in a bureau drawer. It awaits *your* pleasure. This is even true of a visitor. He rings the bell or knocks on the door and you still hold the initiative. You can open the door at your leisure, or under certain circumstances you don't even have to answer it. But let that phone ring and all hell breaks loose; in summer and winter, in bed or out of bed, in the bathtub or up on the roof you make a beeline for that instrument, over hill and dale, in the darkness with the furniture falling to the left and the right; nothing matters except to reach that instrument; and then what? A wrong number perhaps, or some fellow says, "How are things?"

—*Harry Golden,* For 2¢ Plain, *1959*

JFK ship to shore, aboard USS Observation Island, *November 16, 1963.*

If I'm going to propose, we might as well have a good line.

—President Bill Clinton, suffering a fuzzy phone connection as he attempts to offer a spot on the Supreme Court bench to Ruth Bader Ginsburg, 1993

Hello? Er, hello, Di—hello, Dimitri. Listen, I—I can't hear too well. Do you suppose you could turn the music down just a little? Ah, ah, that's much better. Yes. Fine. I can hear you now, Dimitri, clear and plain and coming through fine. I'm coming through fine, too, eh? Good. Then—well, then, as you say, we're both coming through fine. Good. Well, it's good that you're fine then, and—and I'm fine. I agree with you, it's great to be fine. Now then, Dimitri, you know how we've always talked about the possibility of something going wrong with the bomb. The bomb, Dimitri. The *hydrogen* bomb.

—Peter Sellers, in Columbia's Dr. Strangelove, or: How I Learned To Stop Worrying and Love the Bomb, *1963*

The Watergate Connection

That evening, the *Post's* night city editor called Woodward at home. The *Los Angeles Times* was predicting on its front page that the White House would make a dramatic Watergate admission in a few days: one or more high-level officials not identified in the story would be named as directing or condoning political espionage and sabotage activities without approval from the President.

Woodward made an emergency call to Deep Throat. The procedure involved making a call from a pre-designated phone booth, saying nothing and then hanging up after 10 seconds. Woodward had to wait for almost an hour by the phone booth before the call was returned.

No meeting was possible that night, Deep Throat said. "You don't have to tell me why you called."

The whole town is going crazy, what's going on? Woodward asked.

"You'd better hang on for this," said Deep Throat. "Dean and Haldeman are out—for sure."

Out? Woodward repeated, dumbfounded.

"Out. They'll resign. There's no way the President can avoid it."

Could the *Post* publish that?

"Yes. It's solid," Deep Throat said.

What should we do? Woodward asked.

"Someone's talking. Several are talking—go find out. I've got to go. I mean it—find out." Deep Throat hung up.

—*Carl Bernstein and Bob Woodward,* All the President's Men, *1974*

Dustin Hoffman as Woodward's cohort Bernstein in Warner Bros.' 1976 movie version.

The Five Most Irritating Secretarial Queries

Whom shall I say is calling?

Will she know what this is in reference to?

And you are with—?

Can I tell him what it's about?

Does he know you?

Dolly Parton fields a call in Twentieth Century-Fox's Nine to Five *(1980).*

The Seven Deadliest Excuses

He's in conference.

She's just stepped away from her desk.

She's just gone into a meeting.

He's on another call and has three calls waiting.

She's out of town until Tuesday.

He's with a patient.

She's in court today.

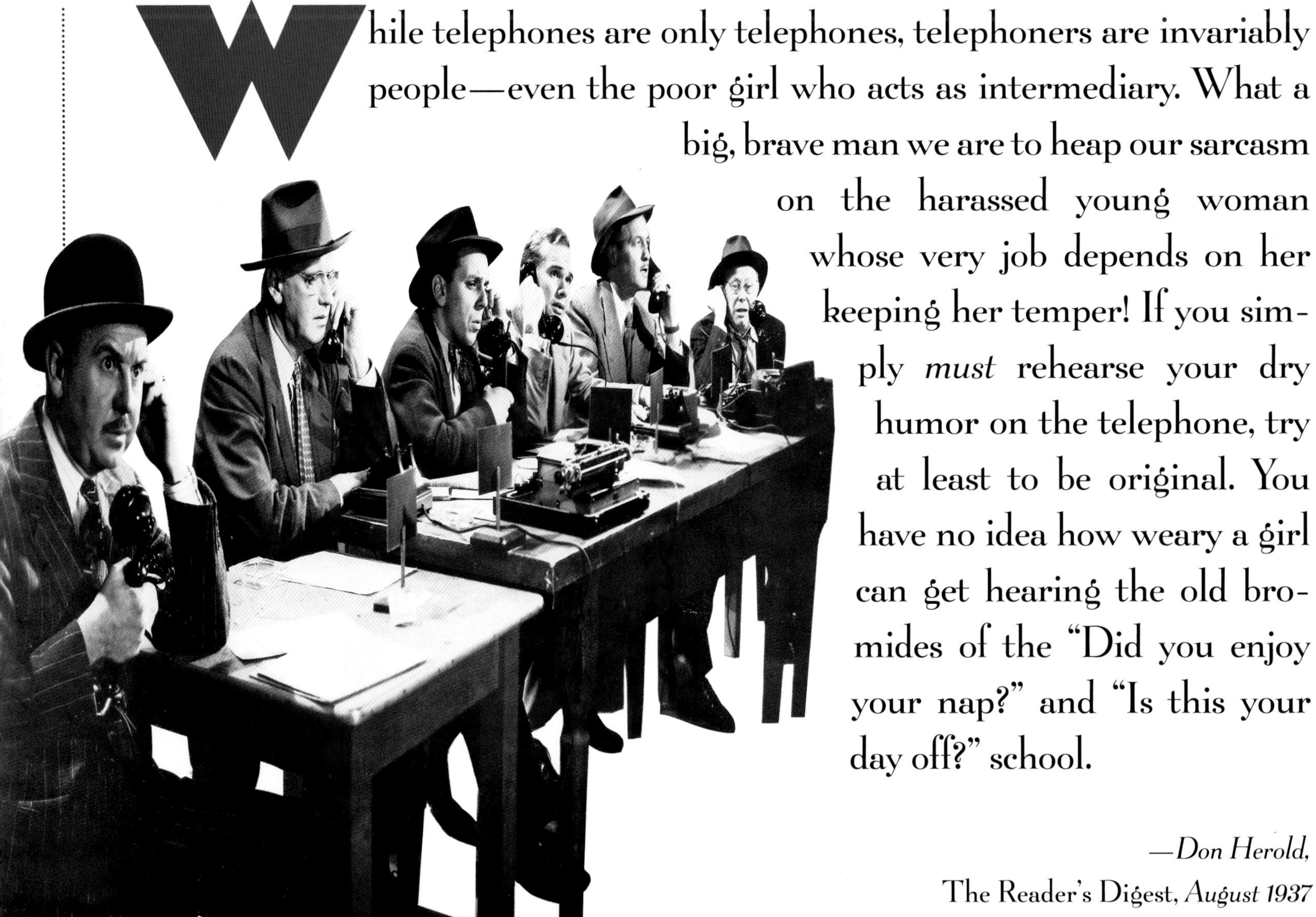

While telephones are only telephones, telephoners are invariably people—even the poor girl who acts as intermediary. What a big, brave man we are to heap our sarcasm on the harassed young woman whose very job depends on her keeping her temper! If you simply *must* rehearse your dry humor on the telephone, try at least to be original. You have no idea how weary a girl can get hearing the old bromides of the "Did you enjoy your nap?" and "Is this your day off?" school.

—*Don Herold,*
The Reader's Digest, *August 1937*

Another essential in the impression you are making is your voice. Its tones, its inflections, go out to everyone in the room, to every customer to whom you speak, to every unseen prospective customer you answer on the telephone. A girl's chances can be ruined immediately by her voice.

In one office there were two secretaries and a very observant and clever employer. In the beginning there was no special girl to answer the telephone. First one girl answered and then the other. Soon the employer noticed that when Miss Carmichael was at the telephone he got the order, but when Miss Milton answered he lost the order. He made it his business to find out the difference. He saw that Miss Carmichael, when answering, seemed to be pleased to have a little respite from her regular work. She looked really animated, as if she knew the person talking at the other end of the phone and liked him. Miss Milton looked cross. She didn't like to be interrupted. She sounded annoyed and curt.

In addition, one day he heard Miss Carmichael say, "No, Mr. Jackson is not in the office at present. He is out on a special case. May I take a message?" and a little later that day he heard Miss Milton say, "No, Mr. Jackson is not in. I don't know where he is. I don't know when he will be back. Good-bye."

Miss Milton was a good worker, but she might prove expensive to the office, for she lost customers. However, the employer gave her another chance, though she was not allowed to answer the telephone. Miss Carmichael took all telephone messages and at the end of the year she was given a bonus for the orders she had obtained.

The telephone voice is not merely the sound of the voice, its musical quality or its harshness, but it is made up, too, of the words that are spoken. Words are not just sounds; words are meant to accomplish something. The successful girl learns when the words she speaks may be her own words and when the words must be those of the company employing her. She is then not her whole rounded self but her office self; she is wearing her office personality. Her words should represent the company.

—Gulielma Fell Alsop, M.D., and May F. McBride, She's Off to Work, *1941*

(Opposite) *Getting the news in* The Sun Sets at Dawn *(1950).* (Above) *A bevy of 1941 beauties in Twentieth Century-Fox's* That Night in Rio.

The ring shatters the night. It jars you out of sleep, but leaves you struggling in the limbo short of full wakefulness.

It rings again. As you fumble for the receiver, the room takes shape around you. By the light reflected from the street, you can make out where you are—in your own home, in your own bed, safe and warm. You relax, savoring security.

The phone rings a third time. The tone is full of menace. The false promise of safety is dispelled and you remember that you are alone and very vulnerable.

The fourth ring is harsh and demands an answer. Hand shaking, you pick up the receiver.

Will he speak? Will he spew out a string of invectives, sly and ugly innuendoes, obscene suggestions? Or will he choose to be silent, forcing you to be the one to speak?

You want to hang up and put an end to it, but you know he'll only call again and keep calling through the darkness into the dawn.

—Lillian O'Donnell, mystery writer, 1993

In the tangled networks of a great city the telephone is the unseen line between a million lives. It is the servant of our common needs—the confidante of our inmost secrets . . . life and happiness wait upon its ring . . . and horror . . . and loneliness . . . and *death!!!*

—*opening crawl from Paramount's* Sorry, Wrong Number, *1948*

Dialing and Dealing

Hollywood runs on information. Who learns what first is important. If it's in *Variety,* it's already too late. The instrument of information is the telephone. Whoever wields the phone most effectively is the winner. Agents, producers, and executives treat their phone as an object of love, cooing into it, making it sing, as if the phone itself were the information sought. Every morning, all over town, people ring up friends and enemies with some minor variant of "Hi, how are you? What have you heard?"

The social and business rituals that have grown up around the use of the telephone are almost as vital as the device itself. If an aspiring agent shows no instinct for the phone rituals of Hollywood, then you're looking at a future former agent. No one explains it, but the quick ones soon learn that a junior agent may call his equivalent across town without an invitation or purpose larger than the exchange of information. But that junior agent may not call a senior agent at another agency without a specific, easily stated, purpose. And even then, a ballet of call-backs and messages ensues. Anyone placing a call between 12:45 and 2:30 (lunchtime) is saying, This call isn't important enough to be made during prime hours. Calls that count are placed between 10 and noon and 2:45 and 4:00.

Calls placed from car phones are less important than calls made from offices or homes. Partly that's because the technology is so lousy that the calls are neither secure nor clearly audible. They seem to imply a casualness that is patronizing.

Agents are by far the most telephone-besotted callers in Hollywood. They often receive more than two hundred calls a day and often place as many themselves. They learn early on that not all calls need be returned. Some are just ignored until the caller goes away. Some of the agents enjoy this, as a form of sadism. Most, however, recognize that being an agent isn't a good way to win a popularity contest, nor is it a profession that turns on good manners. Agents pursue calls that are in their interest. They ignore the rest.

Some agents seem to have a physical connection to their phones. One young agent was seen on a PBS documentary about Hollywood wrapping himself in the telephone cord as he spoke. For a moment or two he appeared in danger of strangulation. He was so involved in his conversation he didn't seem to notice.

A few years ago, during an electrical outage in West Hollywood, the telephones at ICM went out. People who were there said that all the agents looked first stunned, then shattered. They didn't know what to do with themselves. Then, as if by signal, they all grabbed their phone sheets, hurried down to the garage, and sat in their cars calling away until the power returned.

—David Freeman, novelist and screenwriter, 1993

Awaiting the big call in American International's 1958 Attack of the Puppet People.

I'm at work at eight o'clock in the morning. It's afternoon in Europe, for Christ's sake, you've got to be. You never get your telephoning done otherwise. I can call London, the operator's number is 11348, I just dial her and give her the number, I'm connected in three, four minutes. You dial Rome, the goddamn Italians make you wait an hour sometimes. Then I'm on the telex six or seven times a day to Dick, when it's afternoon here, it's morning out there, and if I want to talk to him about something special, I just pick up the phone and call him.

—Darryl F. Zanuck in The Studio, *by John Gregory Dunne, 1969*

He has a reputation for being extremely dilatory about returning phone calls. As a working method, this is defter than the "Don't call us. We'll call you" approach. Cohn's motto seems more like "Go ahead. Call me. I dare you." Filed somewhere in his memory is an invisible, protean "A" list—certain clients, certain people in the theatre, certain people at certain motion-picture studios at certain times—made up of those who, if they were to call and not get through to Cohn immediately, could expect him to call back within, say, twenty-four hours. Otherwise, he has made the unrequited phone call an art.

—Mark Singer, on agent Sam Cohn, in "Mr. Personality", 1988

(Above) *Les Tremayne and Claudia Morgan as radio's Nick and Nora Charles in* The Thin Man; *Carol Kane in Columbia's* When a Stranger Calls *(1979); Ed Begley, Jr. and Jeff Goldblum in New World Pictures'* Transylvania 6-5000 *(1985)*

Sam is a spellbinder by telephone. If television arrives in time to let him transmit his hypnotic eye and his library of facial expressions, he will be greater than ever. As it is, even with the telephone in a rudimentary state, Sam can project a good deal of his personality over it. Writers have been shanghaied from New York to Hollywood by a few minutes of conversation with Goldwyn. They say "No" at first. Sam talks. Finally, they go as if extradited. . . . Even with his business rivals and business enemies, Sam is a siren over the telephone. Executives in other companies wake up in a daze after a telephone conversation with Sam and dimly remember, to their dismay, that they have promised to lend him stars, directors or technical men. Sam once telephoned to David Selznick and asked to borrow George Cukor, one of the greatest of the directors. "Absolutely not," said Selznick, but he finally broke down and consented. Cukor, however, refused to work for Goldwyn. Then Sam telephoned to Selznick again; he wanted the reluctant Selznick to persuade the reluctant Cukor to direct a picture for Goldwyn. Cukor was firm. Sam finally had to confess defeat, and it was on this occasion that he said, "That's the way with these directors; they're always biting the hand that lays the golden egg."

—Alva Johnston, The Great Goldwyn, *1937*

The Scene is the press room of the Criminal Courts Building, Chicago. The room is octagonal, and the audience sees five sides of the octagon. On the down Right side are two high windows; further up stage, at an angle, are double doors; the rear wall, which runs parallel with the foots, has two tables against it, up Right Center against the wall, and Center sticking out into the room. The first has two telephones and a typewriter on it. The second, one telephone and a pile of papers. The next wall, against an angle, has a huge roll-top desk against it; then comes the final wall, down Left, in which is the door Left to the toilet. A long table is Right Center at an angle. There are chairs around it, and four phones on it. The roll-top desk also has a phone on it. Six of these phones communicate directly with various newspapers; Number 4 phone on large Center table is an outside phone, an extension of the switchboard in the Criminal Courts Building. There is a water cooler between the big desk and the toilet door.

—*Stage directions, Act One. Ben Hecht and Charles MacArthur,* The Front Page, *1928*

I talk over the phone for hours. It is a talent with me. When I am really busy I can keep these conversations fairly short—and must, because the phones ring constantly. I love to work under high pressure, this fact probably helping to keep my nerves from short-circuiting. The people who phone me during working hours range from my personal friends to avowed enemies, from conniving rats to purring cats.

—*Louella Parsons,* The Gay Illiterate, *1944*

On the Telephone

The telephone rang the other day. A small feminine voice asked me if I was I, and being assured said, "Just a minute, please, Mr. Gundelfinger wishes to speak to you."

That was not the name, but it will do.

Two minutes passed. I hung on. Then the lady returned, asked me who I was again, and told me Mr. Gundelfinger wished to speak to me. By this time I was wondering who Mr. Gundelfinger was.

The lady went away and there was another silence, and I kept wondering. Then a male voice came on the phone and I thought "Now we'll find out who Gundelfinger is."

But it wasn't Mr. Gundelfinger at all. It was a secretary. He said to hold on please Mr. Gundelfinger wished to speak to me. He said it in a hushed, awed tone; the tone an equerry might use in telling me the Queen desired me to pop around to the Palace for tea; the tone a Hollywood press agent might use in telling me Miss Hedy Lamarr wished to marry me.

I got it. "Mr. Gundelfinger" was really Pr-id-nt Tr-m-n calling to ask if I could become his unofficial envoy to Russia, and for reasons best known to himself, wishing to remain incog.

The secretary went away and left me in a state of high excitement. There was another long silence. Then a brisk, important voice rushed on the wire and said "GUNDELFINGER SPEAKING."

It was a rather pompous voice. I felt disappointed. But Gundelfinger's tone indicated plainly that I was supposed to know who he was, without further clues from him. And I didn't.

Well, it turned out there was no reason why I should. Gundelfinger proved to be a total stranger and all the fanfare, build-up and whoop-de-do prefaced a totally unimportant call. I was to be given the honor of doing something for Gundelfinger for nothing.

I fear I was rather brisk with him, after I got his number. Later I regretted this. I am now sure that poor Gundelfinger happened to be in the doldrums at the time he made that call and needed its fillip to make himself feel important again.

Every man has moments when dark and horrid suspicions of his own unimportance disturb his soul; when people won't take him at his own valuation. At such times reinflation of the ego is vitally necessary. And there are a great many Americans for whom a telephone call such as the one I have just described serves that purpose.

It made Mr. Gundelfinger feel important to get that telephone girl and that secretary revolving around his ego, and getting me to hold onto the other end of the wire for five minutes. It would have made him feel more important had I co-operated.

I wish I were more of a telephone user. I use it only for emergencies, such as calling the doctor, reporting a fire or burglary and asking room service to send in cracked ice, glasses and soda. When in disgrace with fortune and men's eyes, I should like to call up people, keep them waiting several minutes so that they'd appreciate that I am a hell of an important fellow, and then come on the phone with my best director's voice and say, "SULLIVAN SPEAKING!" Only I'm afraid they'd say "Who?" and give me the same brush-off I have Mr. Gundelfinger.

—*Frank Sullivan,* A Rock in Every Snowball, *1946*

Mickey Rooney double-talks in Andy Hardy's Private Secretary *from MGM.*

The Royal Telephone (Sacred Song)

Central's never busy
Always on the line,
You can hear from Heaven
Almost any time;
'Tis a Royal Service,
Free for One and All,
When you get in trouble,
Give this Royal Line a call.

—Chas. H. Powell and Peter Shupe, 1914

Dennis Cardinal Dougherty, the Archbishop of Philadelphia, waiting for a call from a bishop in the Philippines, 1935.

Dreaming . . .

Of a telephone:
Your curiosity will be satisfied.

Making a telephone call:
Advantages in business.

Receiving a telephone call:
Postponement of a date.

Talking long distance on the telephone:
Happiness.

Being without a telephone:
Desires will be realized.

—Zolar's Encyclopedia and
Dictionary of Dreams, *1963*

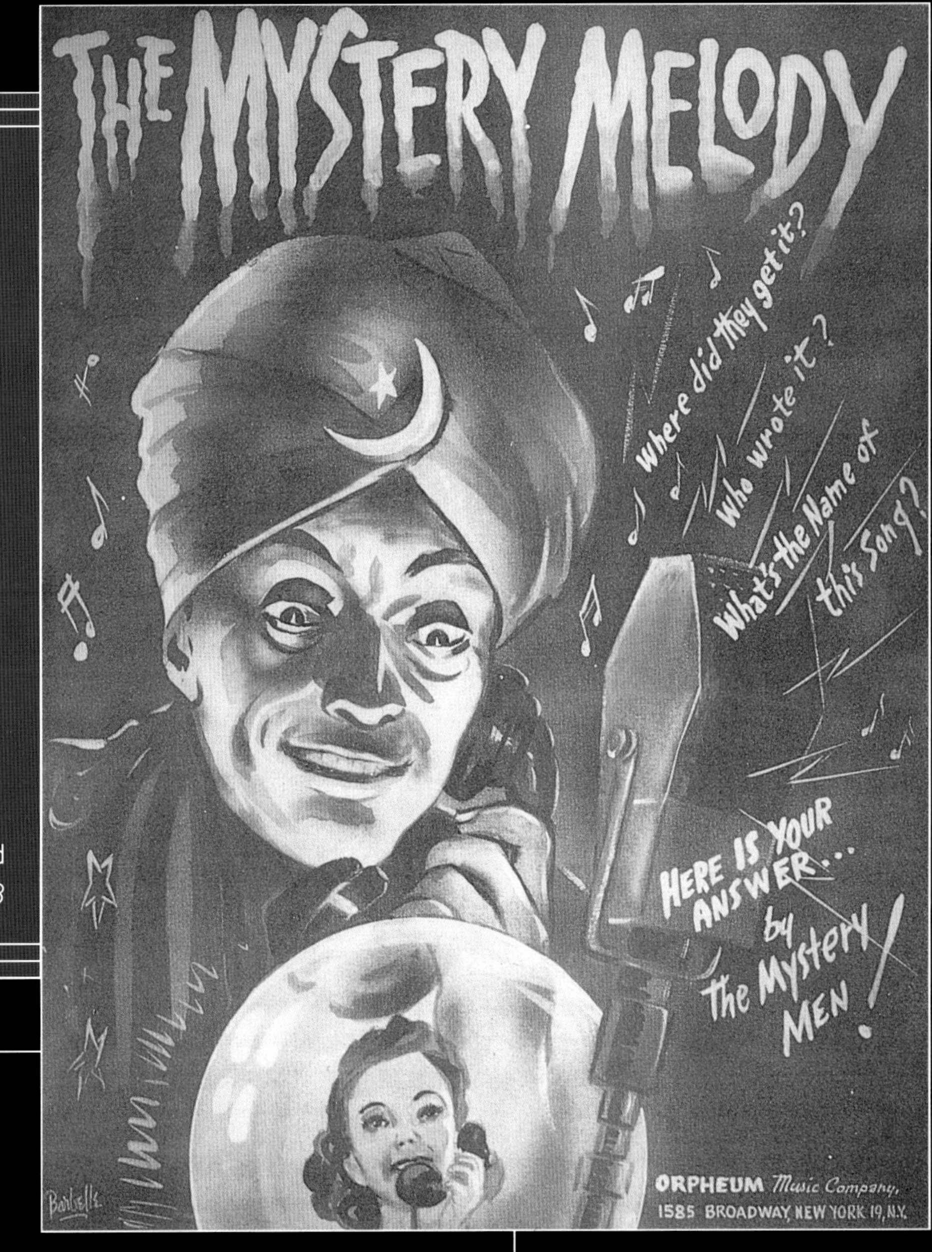

Little Black Heart of the Telephone

That telephone keeps screaming its little black heart out:
Nobody there? Oh, nobody's there!—and the blank room bleeds
For the poor little black bleeding heart of the telephone.
I, too, have suffered. I know how it feels
When you scream and scream, and nobody's there.
I am feeling that way this goddam minute,
If for no particular reason.

—*Robert Penn Warren*

THE SATURDAY EVENING

POST

NOVEMBER 19, 1949 15¢

HOUSEKEEPING HEADACHES IN A POLICE STATE

By Lt. Gen. Walter Bedell Smith

FORMER U. S. AMBASSADOR TO RUSSIA

DUBINSKY:

DICTATOR IN SHEEP'S CLOTHING

Chapter 8

Gigglegiggle, Talktalktalk

Adults take the telephone for granted, but little children sense the wonder. How strange it must look from a child's-eye view. This funny-looking machine is sitting on a table. It makes a loud, insistent, rhythmic, repetitive noise unlike any other. And everyone reacts. Daddy dashes in from the garden, Mommy from her pot of stew. He or she lifts one part of it off another part, places it next to an ear, says "Hello?" and is suddenly talking to it, laughing at it, nodding, smiling, frowning. How odd! They don't talk to the microwave or the lawn mower this way.

The small child is as perplexed as the skeptic of a century ago, especially when asked, "Do you want to speak to Grandma?" Yes, of course, but where is she? Soon the child accepts the strange concept that she's

Telephone Courtesy

Then the telephone. Children usually love to use it and they should be taught to speak courteously on the pain of not being allowed to answer it. Children commit all sorts of discourtesies over the telephone if not checked and one often hears the casual "Yep" and "What" and "Wait."

—*Elsie C. Mead and Theodora Mead Abel, Ph.D.*
Good Manners for Children, *1926*

Faults That the Family Must Be Trained Out of

Question-asking is certainly one of them. "Where have you been?" "Where are you going?" "Why did you do this?" "What are you going to do about that?" "Where, why, why, where?" everlastingly.

For instance, Mamma is sweet as can be, but never NEVER have I been called to the telephone without her asking "Who was that?" "What did he (or she) call you up about?"

It's just a habit. She really doesn't care what I say as long as my voice answers something. But other girls' mothers keep on asking question after question and to each they *want an answer.* They may get a true one or otherwise—but an answer they must have. Grandmother is the one exception. Maybe Bill trained her when he was little. Anyway, she's trained. When I'm in her room and the telephone rings for me, nine chances to ten she will put the radio earphones over her head rather than overhear my end. Let Families take a lesson from that!

I suppose you're waiting to hear me add sweetly: "And so I tell the darling everything!" Which would be perfectly idiotic. But if I happen to feel like telling her things, I do. And I certainly feel like it more often than if I were continually asked to! Mabel's curiosity is simply a mania! In fact, she is not above listening in on another extension. That's how I first managed to have a private telephone in Bill's house. I heard the click of the extension being lifted off as I was talking to Kay, so I said, "Don't tell anyone what you told me about my stepmother," and then broke into a language we talked at Gardenton, to explain. Mabel's curiosity could not resist asking me what was said about her. So I went to Bill and I got my telephone.

Telephones ought to be classed with toothbrushes and each member of the Family have sole right to his own. I know the trouble my friends have waiting, and waiting, while a market maniac calls for tips on every stock in Wall Street, or while aunt or grandmother has one of those "isn't it terrible" conversations with her dearest friend about all their other friends who seem to meet perpetual calamity.

I really don't know whether it is a pet economy idea like saving string and wrapping-paper, or whether it is an idea that a young girl with a private telephone must want it for purposes of intrigue.

Of course, there is nothing we would adore so much as to be really intriguing if we could only learn how. But it is a special gift, something like personality and charm, and imitations are not very good.

—*Emily Post,* How To Behave—Though a Debutante, *1928*

In contemporary life the mastery of the telephone is a proof of the approach of age. Like the first tooth, the last diaper, and the formula no longer needed; like those great moments when rolling over is transformed into crawling and crawling into perilous steps; like those releasing days when shoelaces and neckties can at last be tied and handkerchiefs used with accuracy; like those genuine occasions when the scooter succeeds the velocipede and the bicycle ousts the scooter; when parents' freezing arms are replaced by water-wings and waterwings by breast strokes; or when playschool turns into dayschool, and short pants into long, the full uninhibited employment of the telephone comes as a milepost on the difficult path to growing up.

"He is very good at the telephone," we say of a seven-year-old, meaning that he can take messages with as much accuracy as the operator of a hotel switchboard. Although this may be true, all things considered it is the most niggardly of praise. The stubborn fact is, however, that the children we go to such pains to initiate soon take over.

—*John Mason Brown,* The Saturday Review, *June 19, 1948*

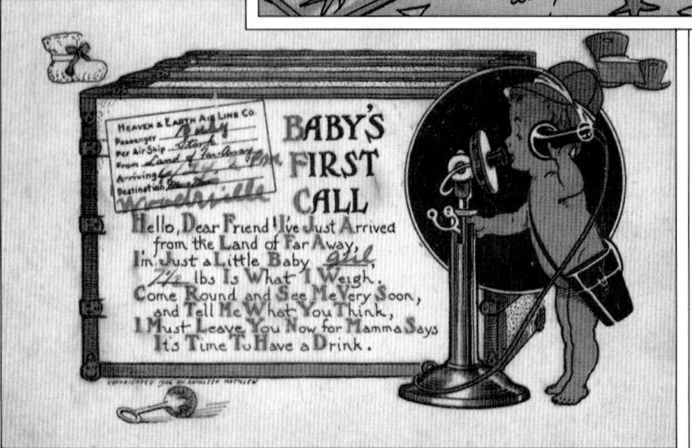

somewhere out there—at the other end of the telephone wire—even if she can't be seen. And not only can she call him, she can *be* called if tiny fingers push the right buttons or dial the right numbers.

Mimicry begets growth, and soon a toy telephone becomes an essential tool. Using a play dial phone or a newfangled "talking" one, the child can call imaginary friends, make imaginary appointments, alert an imaginary fire department. Once telephony is understood, the child can expand his or her repertoire from relatives to rescuers, from Auntie Em to 911.

But grasping the principles does not confer total control. For all its allure, the telephone can also be heartless and exclusionary. What is a merry jingle for

I have a telephone phobia! I was born in Yokohama, Japan, in 1903. In those days my mother sent messages to friends by "chit." A ricksha coolie carried the chitbook and brought back a response.

When a telephone was introduced in our household, it was regarded warily. The operator responded with the phrase, "Moshi-moshi," repeated over and over. I never have known what it means. If anything really happened and a connection was made, it was regarded as a miracle.

When we moved to Manila there were more telephones, though I never used one. I was fifteen when we came to the States, and telephones were everywhere—but still instruments of terror to me. I have overcome some of this in talking to family and friends, but I never "chat." And I will go to any lengths to avoid calling up strangers.

—Phyllis A. Whitney, mystery writer, 1993

I remember the time in the 1930s when I was about twelve and the telephone link with my classmate and best friend, Gloria, was its strongest. We didn't even have to speak.

Once we were both sick in bed in our houses, which were diagonally across the street from one another in a beachside suburb in Queens, New York. To me, sick in bed meant a new book of paper dolls and Social Tea cookies and the daytime, daily radio serials: "Our Gal Sunday," "Helen Trent," "Ma Perkins," and more. Gloria and I got on the phone together, each lying in her own sickbed, to listen to the serials together, the phone receiver to the ear, saying not a word for an hour and a half. Anybody wanting our mothers was out of luck, which is the case whether the teenager is speaking or not.

—Elaine Greene, editor, 1993

Preteen Talk

Q: How do you feel about the telephone?

Stephanie: I love it, because I love talking to my friends, and that's the time to gossip if you don't do it at school.

Katie: At school the teachers will kill you if you whisper to people, so you can't really gossip unless you do it at recess.

S: You want recess to be some time when you can, like, get exercise and play around.

Q: What do you talk about on the phone?

K: You talk about that day. You talk about if the teachers did anything mean, or if some kid did something wrong or something.

S: Boys.

Q: What about boys?

S: If you like them or something.

K: Or if some girl has a crush on someone.

Q: How many calls do you make every night?

K: I get more than I make. I haven't lived through one day without Stephanie calling me or me calling Stephanie. We call each other every single day.

S: We do.

Q: For how long?

S: Depends.

K: It can be really like an hour, to five minutes. We were on the phone one time, me and Stephanie on extensions, and a boy from school. We talked for about an hour and a half, not noticing, just gossiping and gossiping.

Q: Do you have your own phone?

S: No. My parents think I'd have too big a phone bill.

Q: What would you do without the phone?

K: We wouldn't need it because we wouldn't know what it was, but if they took it away suddenly, we'd miss it. And if you write a letter, that's not the same thing, because it won't get there in time.

—Katie Stern, 9, and Stephanie Bauman, 10

the child is a siren for Mom. It rings and she runs. It interrupts them and then engages her completely. It may be a plaything they share, but it is also one smooth-talking rival.

"Most children *hate* it when their mother is on the telephone, especially when her conversations run on and on," according to child psychologists Louise Bates Ames and Frances Ilg in their estimable guide to *Your Four Year Old*. "They hate it; they behave badly and get into trouble and then make a fuss; and then their mother scolds and yells and punishes. It is probably hard for a mother who may feel that she has little enough fun or time to herself during the day to appreciate how shut out a child feels when his mother is gossiping on the phone. It would be an act of great kindness, and would prevent much bad behavior and the accompanying need for punishment, if long, chatty phone calls could be restricted to some time when your child is having his nap or is otherwise out of the room."

From four to eight, interest flags. The child is not yet involved in group socialization and hasn't yet realized precisely how the phone can serve him. But at eight or nine, bonding begins—and the phone becomes a beacon. At ten, the young gossip is tying up the line with after-school chatter. At twelve, the kid is making prank calls and ordering pizza for the family. At fourteen, the chatterer has an extension and is clogging the line. And so it goes, with the eventual acquisition, perhaps, of one's very own phone and very own number. What one does with that becomes one's very own business.

What's the obsession? Why do they rush home from school to phone friends they've just spent the

Hello Central! Give me 510
Ring off. The line is busy

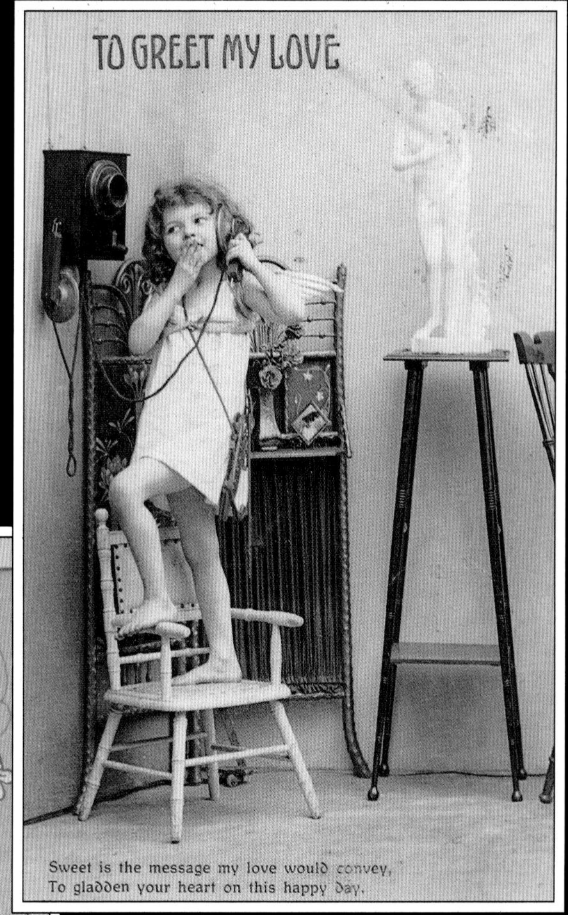
TO GREET MY LOVE
Sweet is the message my love would convey,
To gladden your heart on this happy day.

Just called to wish you a...
HAPPY BIRTHDAY
"The blessing of the Lord be upon you."—Psalm 129:8

10¢ JR. YANK'S 10¢
WAWKY-TAWKY
STRING PHONE
IT REALLY WORKS
5
MINUTES TO MAKE
LOADS OF FUN
DIAPHRAGM MATERIAL
CARDBOARD
30 FEET OF LINE
ADHESIVE PAPER
COMPLETE ASSEMBLING INSTRUCTIONS

As the composer of "The Telephone Hour" from *Bye Bye Birdie,* I was, in 1960, happily unaware of the total intensity, the life-threatening necessity (not to mention the trauma of the telephone bill) of the telephone to the American teenager. It was merely a lark: the prom . . . can I study at your house? . . . what are you doing Saturday?

I was single at that time. Since then I have married, had four children; the conference and hold buttons have been invented; and meals (and homework) have become simply irritating distractions from the main business of life, telephoning.

If I had anything to do with this ("The Telephone Hour" was meant to be an entertainment, a teeny part of life), I'd like to apologize to parents everywhere for my deed.

I can be reached at Debtors' Prison, New York City (area code 212 if you can get to a phone).

—*Charles Strouse,*
composer,
1993

He Says, She Says

"Well," Zock said. "I guess you'll just have to call her up, Euripides."

"Ho, ho, ho," I said. "I'm not going to do it."

"Suit yourself," Zock told me.

"What would I say to her, Zock? I could never think of anything to say to her."

"Try it now," Zock said.

I took a deep breath. "Hello, Sally," I tried. "This here is Ray Trevitt. What the hell are you doing on Saturday?"

"I'm afraid that's not quite it," Bunny said.

"A trifle blunt," Zock agreed, shaking his head.

"Well. It's the best I can do."

"Women have to be coaxed, Ripper. They're funny about that. But you have to play their rules or they'll pick up their baseball and go home."

"Zock," I said. "Help me."

"All right." He nodded. "It's the least I can do. We'll call her this afternoon. At five o'clock."

"What if she's not home?"

"Bunny will pay her a visit at half past four. Right?" Bunny nodded. "So she'll be there." And with that they both disappeared into his house, leaving me alone.

I didn't see him again until late that afternoon. He came in grinning, waving some sheets of paper.

"I've got it right here," he said.

"You've got what right where?"

He shook the papers in my face. "Here. Here in my hands at this very moment is a copy of the conversation you are going to have with Sally Farmer. All you have to do is read it."

I grabbed the papers. It was just what he had said, a conversation, written like a play with two parts, marked Sally and Euripides.

"Better run through it first," Zock said.

"Well, I don't know," I answered.

"You've nothing to lose," he told me. "So begin at the top."

I sighed, took a deep breath, and started reading.

"Hello," I read. "Is this Sally Farmer?"

"Yes," Zock replied, his voice very high. "This is she."

"Well, this is Ray Trevitt."

"Oh, hello Ray," Zock said. "I'm so glad you called."

"How the hell do you know she's going to say that?" I asked Zock.

"It's only polite," he answered. "And Sally's a very polite girl. Bunny says so. She helped me write this."

"O.K.," I said and went on reading. "I heard you were back from camp and I thought I'd just ring up to say hello."

"That was awfully considerate of you, Ray," Zock said.

"You have a good time at camp this year? I understand you were a junior counselor."

"That's right," Zock said. "I had four seven-year-olds in my cabin."

"Gee," I read. "That sounds like a lot of fun."

"Oh, it was," Zock said. "I loved every minute of it."

"Jesus Christ, Zock," I said, putting down the paper, "this is terrible. She's going to think I'm a moron."

"All right," Zock said, throwing up his hands. "If you don't like it, don't use it. It's no skin off my nose. I don't care. The fact that Bunny and I spent hours writing it shouldn't enter in. If you don't like it, don't use it. For all I care, you can contact her by semaphore."

"I'm sorry," I said. "Please. I'll use it. I'm proud to use it. I'm honored. I just hope it gets better later on."

It did. We went all the way through it, and Zock had actually written a fifteen-minute conversation for the two of us, it took that long to read. And toward the end it got very clever, especially the asking-out part, which was put in such a way that she couldn't possibly refuse.

"Zocker," I said when we were done, "you're a genius."

"Naturally." He looked at his watch. "It's five o'clock. Make the call."

"Sure thing," I said, and I dialed the number. When the receiver got picked up, I put my finger under the first speech and started reading.

"Hello," I said. "Is this Sally Farmer?"

"No," came the answer. "This here is Ingebord."

"Who's Ingebord?" I whispered to Zock. It beat him. "Well," I ad libbed. "Is Sally there?"

"I'll see," was the reply.

"What if she's not there," I said to Zock. "For chrisakes . . ."

"Hello," came a voice on the other end.

I grabbed up the papers. "Hello," I said, reading away. "Is this Sally Farmer?"

"Yes. Who is this?"

"Well," I read. "This is Ray Trevitt."

"Who?" she asked me.

I panicked. "Zock," I whispered. "She says 'who?' What do I say?"

"Tell her who you are," he whispered back.

"Well," I said again. "This is Ray Trevitt."

"I don't know any Ray Trevitt," she said. "You must have the wrong number."

"Cut the crap, Sally Farmer," I yelled into the phone. Zock smacked his forehead and fell on the floor.

"What did you say?" she asked.

"This here is Ray Trevitt," I answered, trying to get calm. "You know. Ray Trevitt."

"Oh, yes," she said, sounding very haughty. "Perhaps I remember."

"You must have the mind of a minnow," I told her. "Seeing as I sat behind you all last year in geometry."

"Oh," she said. "*That* Ray Trevitt."

"The same," I said, starting to read again. "I heard you were back from camp and I thought I'd just ring up to say hello."

"How did you know I was at camp?"

"You have a good time at camp this year? I under—"

"That's really none of your business," she told me.

I went right on reading, mainly because I couldn't think of anything else to do. "I understand you were a junior counselor. Gee. That sounds like a lot of fun."

"What in the world are you talking about?"

"I'm talking," I screamed into the phone, "about your seven lousy four-year-olds. I mean your four lousy seven-year-olds. Sally Farmer," I said, throwing the papers away, "do you know what you can do? You can take—"

"If you called to ask me out," she interrupted, "the answer is no."

"Ho, ho, ho," I said. "Who would want to ask you out anyway? Not me. Not under any conditions."

"In that case," she said, "I accept." Then she hung up.

And so it was arranged.

—*William Goldman,* The Temple of Gold, *1957*

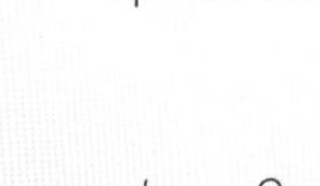

James Ogg in Columbia's Jail Bait *(1955).*

The eager, tireless communication which binds together a coterie of active 14-year-old girls is an impressive manifestation of social psychology. The communication occurs at every available interlude within a day, before and after school hours, whenever two or more girls may meet. When they cannot meet, the telephone brings them together. This is a peak age for interminable phone communications, gay, serious, and hushed. The conversations are punctuated with giggles, gossip, and all sorts of apparent trivia, which, however, are charged with meaning for the young persons on the line.

—*Arnold Gesell, M.D., Frances L. Ilg, M.D., and Louise Bates Ames, Ph.D.,* Youth: The Years from Ten to Sixteen, *1956*

This is a period of transition from childhood to adulthood, a time of painful self-awareness and terrible uncertainty about how to behave in relation to one's peers, as well as to adults, a time filled with so much change, uncertainty and self-doubt that one wants to hide. The mother of a 15-year-old reflected what we see so clearly when she said, "They run in packs to avoid being seen individually, and they avoid daylight so they need not see themselves!"

What better way to "hide" and yet be "involved" than by talking on the telephone! As soon as I began to talk to other parents about it, I found that we were all in agreement about the fact that teenagers often had far more intense and satisfying "relationships" on the phone than they were able to have "in person."

In an earlier age, the first tentative explorations of young people into the realm of adult relationships was probably made via the mail; now the telephone has almost replaced letter writing. Both kinds of communication offer an opportunity to try one's wings—to practice feelings of love and friendship; they are ways of experimenting with new and often frightening feelings. Teenagers are able to say things to each other that they are not yet able to say easily or comfortably face to face. The telephone permits disembodied voices to communicate with each other without the self-consciousness or shyness of being physically real and present for each other.

—*Eda Le Shan*, When Your Child Drives You Crazy, *1985*

entire day with—and, in some sophisticated circles, do this with conference calls? Why is the telephone "treated as if it were a natural appendage of their bodies, an electronic extension of their mouths and ears"?—as psychologist Lawrence Kutner, a columnist for the *New York Times* and the author of *Parent & Child,* puts it. His explanation is that as they move through the amazing new state of self-discovery, kids find their way mostly by talking to people their own age who are going through the same experiences.

The telephone's "unique combination of intimacy and privacy" lets them "test their new skills and new areas of adult behavior without risking as much embarrassment as they would in person," Dr. Kutner adds. "They don't have to be as conscious of their facial expressions and their gestures as they do in person."

What's only this big

But stretches this far?

You use it all the time

Give up? Then turn the page, Mom.

Keystone of the adolescent's communications system is the telephone. If an organization of adolescents were to award a medal for the greatest contribution to life as they know it, Alexander Graham Bell might not win (the inventor of the transistor certainly would give him a run for the money), but he would figure very strongly in the voting. Even the adolescent who is well versed in ancient history and who can conceive of life before electricity, automobiles, and decorator jeans really cannot believe that Julius Caesar did not receive his warning from the soothsayer via Dial-a-Horoscope.

Few adolescents have not at some point requested their own telephone. In part they do so to set a useful precedent for demanding a car of their own. In part they do so because they cannot be sure your phone is not tapped. In the main they do so because with a private telephone an adolescent needs to remove the receiver from its ear only for mealtime, bedtime, and calls of nature, and, with an extra-long extension cord, not even then. If your adolescent has not agitated overmuch for its own private line, that probably is because it knows that in time you will feel the need for a telephone on which you yourself can make and receive calls and that you will run in a new private line of your own.

The telephone also serves as your adolescent's link to the main data bank, far more sophisticated than anything at the Department of the Census. By hooking into the data bank, your adolescent can inform you at any time of what Everybody Else is doing—buying forty-dollar sneakers, going to the skating party, having breast-augmentation surgery.

—*Carol Eisen Rinzler,* Your Adolescent: An Owner's Manual, *1981*

It's your telephone!

Telephone lines often feel like lifelines to me. With friends scattered all over the city, the only way I can contact them with any surety is over the phone. There are competitions: funniest answering-machine message, longest beep (the record is one minute forty-two seconds, with nineteen messages), longest conversation (nine and a half hours), and so on. A separate line has become a necessity; how else can you call a friend at 11:30 P.M.?

There are, of course, difficulties. A few of my friends have to juggle modems, faxes, and the cost of a private line. If you share a line with your parents, Call Waiting is a must, and calls have to be short. Teens and phones however are, together, enough to be the subject of countless jokes, arguments, and sighs of relief as Sally gets through to Sue to give some much needed love-life advice.

Telephones are a way of avoiding the isolation of today's world. When you're in California and your boyfriend is in New York, it's comforting to know that pressing ten buttons will let you hear his voice again. You feel almost as together as you would in the same room. It's cheaper and faster than the trip back, it's easier to ask awkward questions when you're not face-to-face, and you don't have to decide what to wear.

The telephone may only be a substitute for seeing someone in person, but with it you can reach anyone, anywhere, any time (barring parental intervention), and that's what makes it so important.

—Rose Platt, 15

The telephone is a powerful toy—whether it's a grinning make-believe Chatter Phone, a powder-pink Princess, a Mickey, Snoopy, or push-button sneaker. Even with nothing to say, kids love to linger on the line. "By the time they're teenagers, it's a nightmare," says Dr. T. Berry Brazelton, the renowned pediatrician and author (who admits to hating the telephone, necessary as it is in his line of work). "Their ties to each other are so critical. You can't get them out of the shower and you can't get them off the phone."

The Saturday Evening

POST

August 26, 1961 — 15¢

AN EX-CONVICT TELLS

WHY PRISONS FAIL

GLAMOUR TREATMENT FOR THE MENTALLY ILL

Chapter 9

The Pay Phone

It started at home, where families subscribed to telephone service and paid a monthly bill to lease the company's instrument. This phone was off-limits to nonsubscribers, however. And early on, there were plenty of these. How, then, to summon the doctor? The police? The fire department? What would happen if a phoneless neighbor used another's phone? Who paid? How? How much? And what if it was three in the morning?

With problems ranging from bookkeeping to friend keeping, it was essential that telephones be made accessible to all. Thus, the first public pay station in the world went into service on June 1, 1880, in the office of the Connecticut Telephone Company in New Haven (where the first exchange had opened two years earlier). For ten cents, paid to a uniformed attendant, anyone could talk to anyone.

Before long, the attended phone had been retired and replaced by the brilliant invention of William Gray. According to legend, Mr. Gray had been turned away by cold-hearted neighbors when he sought to use their telephone to call a doctor during a family emergency. Determined never to suffer such

100 SHARES

N8349

100

INCORPORATED UNDER THE LAWS OF THE STATE OF CONNECTICUT

THE GRAY MANUFACTURING COMPANY

SEE REVERSE FOR CERTAIN DEFINITIONS

This Certifies that SHEARSON HAMMILL & CO. INCORPORATED is the owner of

ONE HUNDRED

shares of the par value of Five Dollars ($5) each, full paid and non assessable of the Common Capital Stock of The Gray Manufacturing Company transferable on the books of the Company by the holder hereof in person or by duly authorized attorney upon surrender of this certificate properly endorsed. This certificate is not valid until countersigned by the Transfer Agent and registered by the Registrar.

In Witness Whereof, The Gray Manufacturing Company has caused its corporate seal to be hereunto affixed and this certificate to be signed by its duly authorized officers.

NOV 6 1969

FIRST NATIONAL CITY BANK

MANUFACTURERS HANOVER TRUST COMPANY

COMMON

Patricia I. Day

SEAL

PRESIDENT

To avoid the wrath of his Boston landlady, Thomas A. Watson rigged up the first telephone booth in 1877 to muffle his voice. The ingenious assistant to Professor Alexander Graham Bell knew he would have to shout to make himself heard on an experimental call to the inventor in New York. By draping blankets over barrel hoops, he shouted his replies to the Professor without arousing tenants in the boarding house.

From this makeshift "first," indoor and outdoor telephone booths now number some half a million in use throughout the United States. The familiar booths are now as much a part of the American scene as the soda fountain and, like the drugstore lunch, they serve the American desire to do things in a hurry—besides offering telephone privacy in public. In stores, on streets, and dotted along highways these friendly booths mean more convenience for telephoning Americans.

—*Telephone Almanac, 1958*

One large Telephone Company states

TO CALL OPERATOR READ DIRECTIONS

that Prepayment Pay Stations net an extra profit of between $18,000 and $25,000 per year by collecting nickels which would otherwise be lost.

1911, 1912

This amount is but a small portion of the regular pay station revenue.

The Baird Prepayment Pay Station.
No Extra Exchange Equipment.
Automatic in Operation.
No Grounds.

Write for Complete Information

BAIRD ELECTRIC COMPANY
3135 N. HALSTED ST., CHICAGO, ILL.
Pay Stations, Time Recorders, Telephone Meters, Peg Counters, Lockout Telephone Systems, and Wire Chief's Testing Equipment

NICKEL PHONE
PURCHASE SLUGS
FROM
CASHIER

dependency again—and thinking also of other phoneless people like himself—he came up with a remedy that would change calling habits forever. The first "coin-controlled apparatus for telephones," which required placing a coin in a slot, was granted patent number 408,709 and installed in the Hartford Bank in 1889.

In 1890, ten coin boxes, manufactured by J. H. Bunnell and Company, appeared in New York. The Baird Manufacturing Company came up with a portable post-pay box (the coin to be deposited after the connection was made). This lanternlike instrument, which had a convenient handle, was frequently toted around to tables in fancy restaurants. Soon there were twenty-five different companies turning out variations on the coin-operated theme. They looked like mail boxes and vaults, fire-alarm boxes and casino slot machines.

The Gray Telephone Pay Station Company led the pack. The next challenge was how to determine the denominations of deposited coins. "Gray and other coin box inventors used electrical signals, sounds from buzzers, plucked reeds, and similar arrangements for indicating a coin deposit, but when more than one coin was involved, a complicated lever mechanism, operated by the caller, was usually employed," explains a technical text from Bell Labs. "About 1890, Gray accidentally dropped a coin on a bell and realized that the resultant sound, picked up by the telephone transmitter, could be used as a coin signal using no energy other than that of the falling coin." What ensued was a separate slot-and-chute system for each of the three most common coins. "The nickel, on its way to the cash box, struck a bell once;

Candy Store Kid

The relationship of the sale of Tootsie Rolls, malteds, and cigarettes to Abe Lebenbaum's profit was not unlike that of a dripping faucet on Riverside Drive to that water level of the Hudson River up at Poughkeepsie. The slot machine, on the other hand, was very important. When the collector came around every Saturday morning and unlocked the steel drawer at the bottom, Abe Lebenbaum received one half the nickels that had accumulated during the week. But it was the telephone booth that kept him in business.

The only private phone in the area was in the home of Dr. Weitz on Fourth Street between Avenue D and Avenue C. He needed it for his business. And he could afford it, of course, because in the end his patients paid for it. But it was not available to the residents of the neighborhood. I mean a resident of East Fourth Street could not ring the doorbell of the Weitz brownstone and say to George's father, "I'd like to use your phone, please, here's a jit." A jit was a nickel. Derivation: the fare for a ride on what was then the city's main form of public transportation: the jitney bus. Okay, so Dr. Weitz has been ruled out, but the resident of East Fourth Street still wanted to communicate with someone by telephone. What other avenues were available to him? He could go up to Lesser's drugstore, on the corner of Avenue C and Eighth Street. Mr. Lesser had a phone booth. Or he could use the phone booth in Abe Lebenbaum's candy store.

Very few people did either. Making telephone service available to all on East Fourth Street was not unlike dotting filling stations all over the Garden of Eden. Who needed it? Answer: the Zabriskie sisters. . . .

I would not be surprised, since they were so similar in appearance and I never saw more than one at a time, to be told at this late date that there had been only one Zabriskie sister, a lady who had moved around fast and preferred to be known as three. I would be surprised, but I wouldn't believe it, either. No one person could have handled the traffic that moved through that sixth-floor tenement flat.

This traffic was controlled by the telephone booth in Abe Lebenbaum's candy store. I'm not saying that during my hours on the job, I didn't sell an occasional Tootsie Roll or once in a while whip up a malted. But most of my time was spent running to the phone booth and taking messages for the Zabriskie sisters.

—*Jerome Weidman,* Last Respects, *1971*

the dime hit the bell twice. The quarter followed an entirely different path, striking a so-called 'cathedral gong.'" Thus was the operator apprised of the amount deposited.

Thanks to the intrusion of slugs, wires, and other mischievous tricks devised over time—the most sophisticated being the arrogant use of a battery-operated pocket-size tape recorder which played the bell sounds into the mouthpiece—the standard holes designated for nickels, dimes, and quarters would be phased out, beginning in 1972. They were replaced by a common slot, just as in the beginning.

What with static and crossed wires, a telephone call—especially long distance—was murky and indistinct. It was also an entertainment for the boys at the bar or the gang at the general store. But the caller wasn't particularly keen on having everyone else knowing his business. Privacy was in order.

The first fully enclosed telephone booth, on wheels, was patented in 1883. Then came a series of more elaborate booths from the Telephone Company.

Appointments may be made and conversations held, giving all the advantages of a personal interview.

As described in an 1891 brochure, basic models—with double walls and domed roofs—were available in oak or cherry and cost between $112 and $225. Extra features included a Wilton rug for $3.50 or $6.50, revolving stools with russet leather tops for $2.00, and yellow silk window draperies for $3.00 a pair. These elegant closets were soon a popular attraction at the grand hotels and watering holes of the rich and famous.

Simplified, stripped-down models, designed to be lined up in a row, were featured in Western Electric's 1912 catalog. Outfitted with double-hinged folding doors, fans, lights, stools affixed to the hardwood walls, and shelves with slots beneath them for the local telephone book, these became the standard for about forty years in every lobby and corner drugstore. Later models, made of steel, went outdoors and upwards—to the elevated railroad platforms of the subway system, the park, the filling station, the bus stop, and the highway.

In the 1950s, the accordion door became glass, and privacy was on the way out; no longer could milady kick off her shoes and change her nylons. Then, to ward off vandalism and other less sanitary abuse, the door disappeared entirely. A new kiosk, designed by Bell Labs in association with Henry Dreyfuss & Associates, had acoustically treated walls, a back wall of curved glass, a roof of translucent plastic, and legs. It flopped.

In Paris, to make a telephone call is an event. Here, telephoning is as easy as breathing. Only for a minute has the man in the straw boater abandoned his Coke, as he drops in the coin, dials, speaks in monosyllabic undertones and is back at the counter while the coin tinkles to its resting-place, and a dying flutter of metal denotes that another telephone call has been made.

—*Cecil Beaton,* Cecil Beaton's New York, *1938*

An Idol at the Algonquín

One day when Gable came to lunch, a dignified lady, the wife of a banker and up to this moment apparently normal, choked on her salad and, rushing from the table, spent the rest of her luncheon hour in the telephone-booth phoning all her acquaintances to come quickly if they didn't want to live the rest of their lives in regret.

—*Frank Case,*
Tales of a Wayward Inn, *1938*

(Above, clockwise from upper left) *Woody Allen in Paramount's* Play It Again, Sam *(1972); Mia Farrow in Paramount's* Rosemary's Baby *(1968); James Stewart in Twentieth Century-Fox's* Call Northside 777 *(1948); Tippi Hedren in Universal's* The Birds *(1963).*

The telephone is our burglar's tool, he thought.

Sitting in the phone booth the next morning, he found it impossible to conceive of anyone ever having had an affair before the telephone was invented. This was the assurance and the reassurance which kept them together during the week-long separation. This was the advance scout which checked and double-checked on possible danger, warned of it, prepared for it. This was the single grappling hook which connected two separately revolving worlds from which two people had somehow been stolen and thrown together. The telephone was an absolute necessity.

And so was the loose change, he thought, reaching into his pocket. Hastily, he deposited his dime and dialed.

Because the phone calls were stolen, they had to be speedily inserted into the normal routine of two separate lives. There was no time to cash a dollar bill or a fifty-cent piece, no time to linger at the cash register where a curious neighbor might engage you in conversation and then surmise you cashed your bill to make a phone call. He could think of only two conceivable reasons for using a local public phone booth if you had a phone at home. Either you were calling your wife to decipher an item on the shopping list or you were calling another woman. He could understand a curious neighbor buying the undecipherable item the first time around. He could not picture that neighbor buying the same story twice. So it was essential that he have ready change in his pockets, change that would take him quickly into a store or a filling station or a restaurant and then quickly to the phone booth. Once inside the booth, he could turn his back to the glass doors and make his call anonymously.

Now, as the phone rang on the other end, as he wondered why Maggie did not answer it, he realized he had learned to hoard small change like a miser.

Nickels and dimes, quarters, he collected faithfully, cached them in his jewelry box with his cuff links. He never left the house without an assortment of change in his pockets. He assumed it was the same for any man involved with another woman, and he wondered what would happen on that fictitious day in the future of America when suspicious housewives across the face of the nation decided to hold an unannounced inspection of their husbands' pockets.

"Hello?" the voice said.

"Hi," he answered. The voice sounded almost like Maggie's, but the shading was slightly off. He almost said "Maggie?" and then something stopped him, something warned him of danger, and he said instead, "Who's this?"

"Who's *this*?" the voice asked.

He was sure now that the woman was not Maggie. He said, "This is Fred Purley of Purley Real Estate. May I speak to Mrs. Gallanzi, please?"

"I think you have the wrong number," the voice said.

"Isn't Isabel Gallanzi there?"

"No," she said, "you have the wrong number."

"Oh, excuse me. I'm sorry," he said, and he hung up.

He called back later that day.

"Hello?" the same voice said.

He recognized the voice at once this time. Abruptly, bringing his voice down an octave, he said, "Lemmee talk to Joe."

"Who?"

"Joe. Joey. Lemmee talk to him."

"There's no Joey here," the woman said. "You have the wrong number."

"Argh, goddamnit," Larry said, and he hung up.

—*Evan Hunter,*
Strangers When We Meet, *1958*

Nowadays, since telephone rates have gone up, it costs more to have an affair. Nowadays, too, it's very difficult to find a working pay telephone in any major city; the vandals have been busy. Besides, with modern technology, you can now carry a cellular phone and make your illicit calls that way—if you don't mind the number showing on your phone bill, and if you don't mind someone hitting you on the head and stealing the phone from you. All things considered, it's probably a better idea to remain faithful.

—*Evan Hunter, 1993*

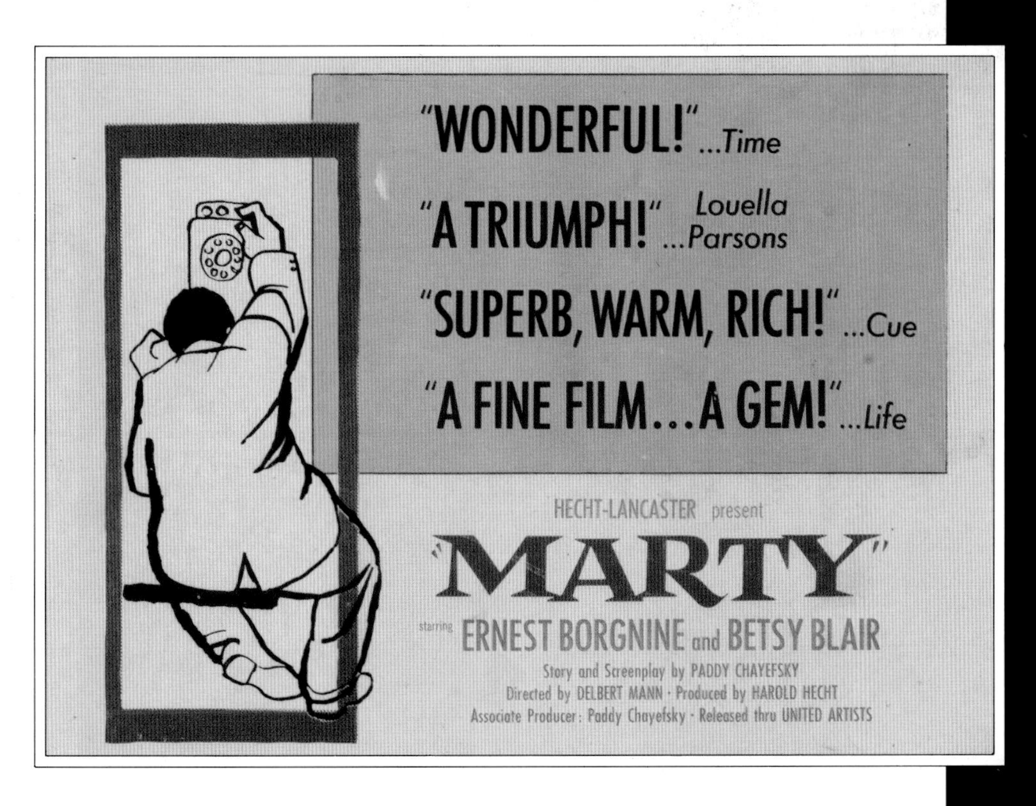

Oh, hello there. Is this Mary Feeney? Hello, there. This is Marty Pilletti. I—I wonder if you recall me. Well, I'm kind of a stocky guy. The last time we met was in the RKO Chester. You was with a friend of yours, and I—I was with a friend of mine, name of Angie. This was about a month ago. The RKO Chester on West Palm Square. Yeah, you was sitting in front of us, and we was annoying you, and—and you got mad and—I'm the fella who works in a butcher shop. Oh, come on, you—you know who I am! That's right, and then—then afterwards we went to Howard Johnson's. We had hamburgers. You hadda milkshake. Yeah, that's right. Yeah, well, I'm the stocky one, the heavy-set fella. Yeah, well, I'm—I'm glad you recall me because I hadda pretty nice time that night, and I was wondering how everything was with you. How's everything? That's swell. Yeah, well, I tell you why I called. I was figuring on taking in a movie tonight, and I was wondering if you and your friend would care to see a movie tonight with me and my friend. Yeah, tonight. Well, I know it's a little late to call for a date, but I didn't know myself till—yeah, I know. Yeah, well, what about—well, how about next Saturday night? Are—are you free next Saturday night? Well, what about the Saturday after that? Yeah. Yeah, I know. Well, I mean, I understand that. Yeah. Yeah.

—Ernest Borgnine, unable to connect, in Marty, *1955*

He squatted down and picked up the dachshund, catching a glimpse of Eddie, the doorman, as he did. Still watching! The dog began bucking and thrashing. Sherman stumbled. He looked down. The leash had gotten wrapped around his legs. He began gimping along the sidewalk. Finally he made it around the corner to the pay telephone. He put the dog down on the sidewalk.

Christ! Almost got away! He grabs the leash just in time. He's sweating. His head is soaked with rain. His heart is pounding. He sticks one arm through the loop in the leash. The dog keeps struggling. The leash is wrapped around Sherman's legs again. He picks up the telephone and cradles it between his shoulder and his ear and fishes around in his pocket for a quarter and drops it in the slot and dials.

Three rings, and a woman's voice: "Hello?"

But it was not Maria's voice. He figured it must be her friend Germaine, the one she sublet the apartment from. So he said: "May I speak to Maria, please?"

The woman said: "Sherman? Is that you?"

Christ! It's Judy! He's dialed his own apartment! He's aghast—paralyzed!

—Tom Wolfe,

The Bonfire of the Vanities, *1987*

To a Lady in a Phone Booth

Plump occupant of Number Eight,
Outside whose door I shift my parcels
And wait and wait and wait and wait
With aching nerves and metatarsals,
I long to comprehend the truth:
What keeps you sitting in that booth?

What compact holds you like a stone?
Whose voice, whose summons rich
with power,
Has fixed you to the telephone
These past three-quarters of an hour?
Can this be love? Or thorns and
prickles?
And where do you get all those
nickels?

—Phyllis McGinley

Eight coin-box telephone booths in the lobby of the Jollity Building serve as offices for promoters and others who cannot raise the price of desk space on an upper floor. The phones are used mostly for incoming calls. It is a matter of perpetual regret to Morty, the renting agent of the building, that he cannot collect rent from the occupants of the booths. He always refers to them as the Telephone Booth Indians, because in their lives the telephone booth furnishes sustenance as well as shelter, as the buffalo did for the Arapahoe and Sioux. A Telephone Booth Indian on the hunt oftens tells a prospective investor to call him at a certain hour in the afternoon, giving the victim the number of the phone in one of the booths. The Indian replies, of course, that it is a private line. Then the Indian has to hang in the booth until the fellow calls. To hang, in Indian language, means to loiter.

—A. J. Liebling, "The Jollity Building," 1942

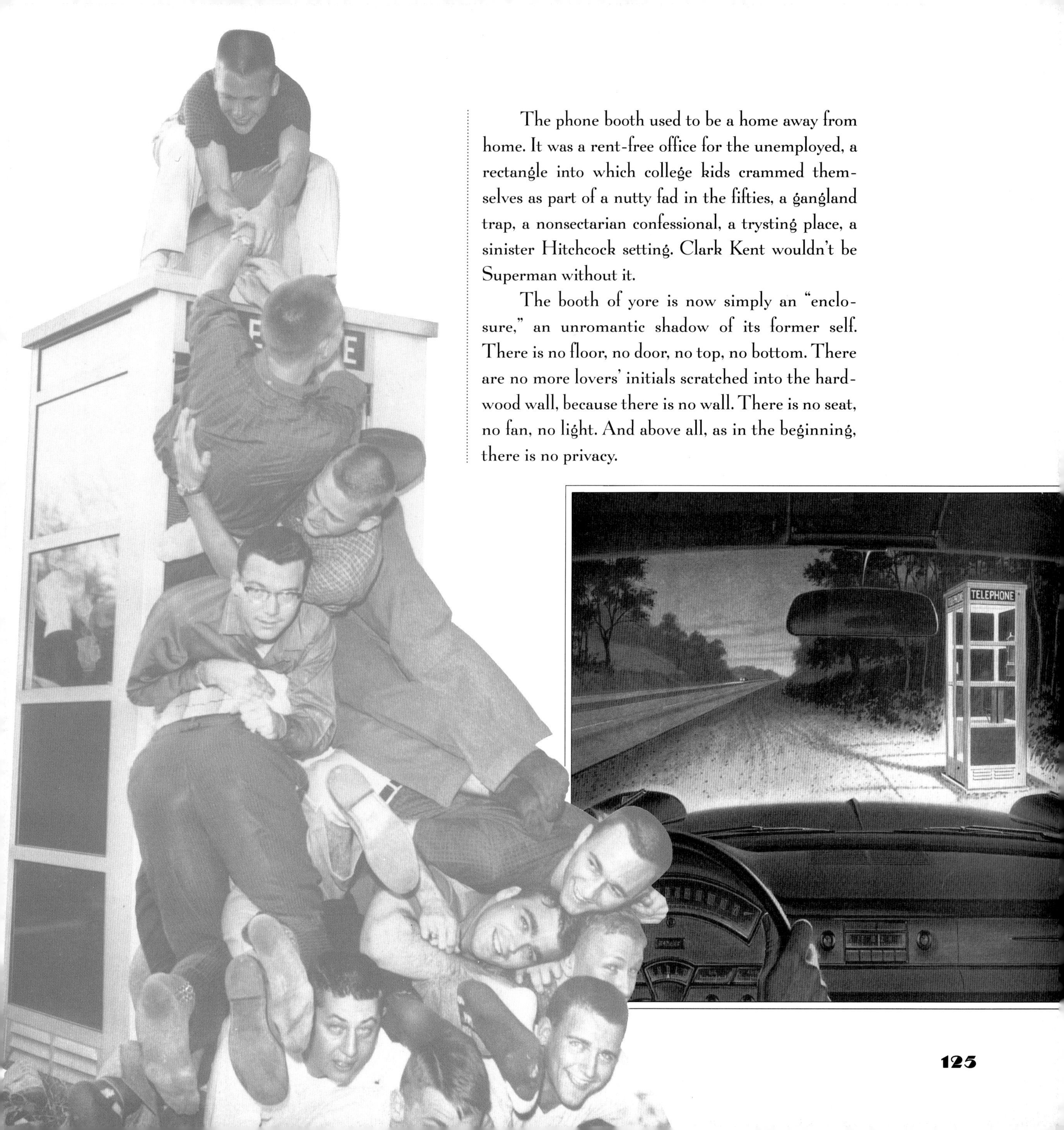

The phone booth used to be a home away from home. It was a rent-free office for the unemployed, a rectangle into which college kids crammed themselves as part of a nutty fad in the fifties, a gangland trap, a nonsectarian confessional, a trysting place, a sinister Hitchcock setting. Clark Kent wouldn't be Superman without it.

The booth of yore is now simply an "enclosure," an unromantic shadow of its former self. There is no floor, no door, no top, no bottom. There are no more lovers' initials scratched into the hardwood wall, because there is no wall. There is no seat, no fan, no light. And above all, as in the beginning, there is no privacy.

Chapter 10

Bell's Hell

In the beginning, the telephone was a miracle and a luxury. Then a convenience. Then a necessity. We were all connected, and we took it for granted. "Ma Bell," the affectionate nickname for AT&T, connected us. In the "Spirit of Service," she cradled her customers, was always on call, and dispatched the telephone repairman in rain and sleet, blizzard and dust storm.

Now we're reconnected, and rewired—not with old-fashioned copper cable wires but with fiber-optic ones. The world's largest corporation—which for a century had a lock on the manufacture of all phones, cables, and communication products through its own Western Electric Company; a lock on all scientific research and product development through its Bell Telephone Laboratories; and a lock on all long-distance service—was forced to disperse its pieces in 1982. With a roll of the dice from the Justice Department, the monopoly game was over.

"But never was a coup more reluctantly accomplished," writes Alvin von Auw, a former AT&T executive, in his reflections on the biggest antitrust suit in history. "What AT&T agreed to abandon was nothing more nor less than its historic mission, providing America with telephone service. It was that

(Opposite) Keith Haring's telephone (1989).
(Above) Earth Station at Andover, ME, the Bell System's ground station for communication by satellites.

Early one evening, I received a telephone call from a computer that asked me about the shampoo I use. I don't usually discuss my toilette with callers who happen to be alive, so I hung up.

As I did, I asked my wife, "Do you remember when the ringing of the phone didn't make us cringe? Do you remember when even callers with wrong numbers were polite?"

I pine for the time when a telephone's bell was a musical prelude to a happy surprise: My father had tickets to a Yankees game; my aunt was sending me an autographed picture of Donna Reed; my teacher had been attacked by bees. . . .

I don't mean to live in the past, but do you remember when call waiting meant waiting for a call? Do you remember the sweet anticipation of a call that might light up your life? Instead, the phone has become an instrument of intimidation.

We're all connected, the telephone company likes to sing, but I don't want to be connected to a computer in Kansas or the Beach Boys in L.A. I want to have a quiet dinner with my wife, without being interrupted by the voice of Visa or Sears.

—*Ralph Schoenstein,* Woman's Day, *March 16, 1993*

mission that gave the 'biggest company on earth' its neighborhood character, that made it one with the communities it served."

Ma Bell's divestiture spawned a litter of baby Bells, just as had happened a hundred years before, when independent and rural phone companies broke from the Bell Company and sprung up across America. In the brave new world of the 1980s, MCI, Sprint, US West, and hundreds of smaller companies finally had access to big-time information, and the industry took off.

But with diversity came disregard. Company soul expired. "Every Tom, Dick, and Harry can open up a phone company—and has," snorted Lily Tomlin's rebellious operator Ernestine over a phone line at New York's Cooper-Hewitt Museum in a 1993 exhibit called *Mechanical Brides.* "There'll be Ma Bell, Pa Bell, Aunt Bell, Cousin Bell, Clara Bell, Baby Bell, Cow Bell, Door Bell, Lulu Bell, Hell's Bells," she grumbled, with an ear to the future. "For all the respect I get, I might as well work for Taco Bell."

Dr. Bell's first clunky box has sped on down the road and onto a microelectronic highway. Call Waiting has replaced the busy signal and provided a new

With every passing day, I am reminded how inadequately equipped I am to confront life as it is lived and evaluated in these times. Take the technology of modern romance. What should one answer when a friend asks at what point in a budding involvement he should take the leap and put his new beloved on autodial? And if he does it on the cellular as well as the home cordless, does that imply a greater degree of commitment? Frankly, m'dear, I haven't the slightest and, moreover, I don't give a damn.

—*Michael M. Thomas, The* New York Observer, *September 27, 1993*

species of complicated social behavior: conversationus interruptus. Call Forwarding electronically redirects the call. When the line is busy, Repeat Dialing beats a redial button, leaving the caller's fingers free to practice the sax or fold the laundry until the connection is electronically made. When a call has been missed by a margin, Call Return will get it back.

The fax machine, hooked up to the phone line, is in many ways a high-tech throwback. It produces an actual piece of paper, reminiscent of the time when letter-writing, not telephone-talking, was the popular form of communication. But more than that, it sends and receives instantaneously—be it a scrawled order for Chinese food, a birthday card to a cross-country colleague, or even, miraculously, an impassioned message to the Wailing Wall in Jerusalem.

TeleTicket means never having to stand in line at the movies, and Reel Review gives a particular film's take on drugs, alcohol, and sex. KidsLine tells

Beep. "Mom, Dad," says a 19-year-old, "I'm sorry, but I'm gay. And I will never, never change." *Beep.* "I just want to say I'm sorry," sobs a young woman, who says she caused an automobile accident that killed five people. "I wish I could bring them back." *Beep.* "I wish I had someone to share this with," murmurs a man, revealing the secret pleasure he gets from wearing his wife's clothing.

The soul baring goes on and on, recorded in 60-second messages to the Apology Sound-Off Line, a Los Angeles-based telephone service that offers the catharsis of confession for the price of a phone call: The service, started up this summer by a Los Angeles outfit called United Communications, receives some 200 anonymous calls a day from people admitting everything from marital infidelity to murder. "They are gut-wrenchingly honest," says apology-line operator M. J. Denton.

—Time, *October 3, 1988*

Ilene Segalove's
Telephone Clutch

how the telephone changes a voice into electric signals. PostalAnswerLine explains the intricacies of parcel post rates, chain-letter fraud, and how dogs can delay the mail. The Psychic Forecast, Dow Jones Market Report, Scorephone, Dial-a-Joke, Dial-a-Prayer, Dial-a-Hearing-Screening-Test, Dial-a-Shrink, Dial-a-Mattress, Dial-a-Pizza, Dial-a-Contact-Lens, Dial-a-Tune, and Latest News Headlines in Korean supply goods and services essential and arcane. Toll-free 800 numbers have hefty phone directories of their own. Costly 900 numbers range from the legitimate to the fraudulent to the obscene.

Wireless and digital signal the future. One in every five new phones is cellular; Americans are buying them at the rate of 7,000 a day. The car phone has become an escape hatch from the traffic jam, enabling the trapped driver to call for help, call ahead, or vent his spleen over talk radio. The folding phone is a rude interruption at lunch or the opera, as if such conversation could not wait. And what with a Dick Tracy-like wrist phone on every obsessive-compulsive's arm and a beeper in every other pocket, no one—no matter how far from home base—is out of touch.

The modem—which modulates data into sound, speeds it across phone lines and then demodulates it back into data—has transformed telecommunications. Modems enable millions of subscribers to order airline tickets, book hotel rooms, shop from hundreds of stores and catalogs, obtain Associated Press stories or Reuters news photos, or correspond with a modem-pal in Moscow through on-line information services such as Prodigy, CompuServe, America On-Line, and Internet.

At the Sound of the Beep . . .

It began just after I'd bought my first telephone answering machine in 1974. Such machines were still somewhat of a luxury back then, but once I installed mine, my next concern was the message.

That morning, I'd interviewed Milton Berle, and on a whim I asked him to do a message for my new machine. Berle was an ironic choice, since forty years before he'd been the first celebrity my father, Broadway columnist Leonard Lyons, had approached for news.

I didn't realize what I'd started until, in short order, I began building up an audio library of interviewees. Each name presented a different challenge. Joe DiMaggio obliged at an Oldtimers' game at Shea Stadium and even asked adoring young fans to be quiet for a minute while he said, "This is Joe DiMaggio. Judy and Jeff aren't here now, so please leave a message."

Jack Nicholson never does television interviews, but after a one-hour session at my radio mike, he was glad to oblige. Muhammad Ali even agreed to recite a poem I'd written for the occasion. Bob Hope did a terse but cordial message. Glenda Jackson showed much enthusiasm, and David Niven tried to be funny. Anne Bancroft sang part of "Yes, Sir, That's My Baby." Mel Brooks did a lengthy imitation of Al Jolson calling Irving Berlin in 1931.

Albert Finney had just released an album, so he inserted a plug, calling himself "the new, hot, pop star," while lyricist Sammy Cahn sang his. James Mason did his Rommel accent from *The Desert Fox*. Peter Ustinov delivered one message in a Texas accent, another in his elegant Queen's English. Peter Sellers did one with a Beatles Liverpool accent, then one imitating Sir Winston Churchill. James Earl Jones used his sinister Darth Vader voice.

Alfred Hitchcock, Alec Guinness, Sophia Loren, Bette Davis, Jeanne Moreau, Michael Caine, Jack Lemmon, Jane Alexander, Sidney Poitier, Lily Tomlin—they all seemed to fall into line, so that today my collection numbers some two hundred actors and actresses. But please don't call just to hear Harrison Ford or Anthony Quinn or Warren Beatty or Shirley MacLaine. They're not on the machine this week. I won't tell you who is, however.

—*Jeffrey Lyons, film critic, 1993*

(Above, left) Warren Beatty in Paramount's Parallax View, *1974.*
(Above, right) Telly Savalas and Sidney Poitier in Paramount's The Slender Thread, *1965.*

A Pressing Matter

I telephoned Taffy's house. Naturally a machine answered. "If it is dinnertime, press 1," said the machine. It was 3:13 P.M., so I did not press 1. The machine said, "If it is not dinnertime, press zero."

I pressed zero and immediately heard an unremitting electronic tone. I had been disconnected.

Immediately I telephoned again. Again I declined to press 1. When the machine again told me to press zero if it was not dinnertime, I pressed zero. Again I was disconnected.

I phoned a third time. Hoping to get around Taffy's machine, I instantly pressed 1 when it said, "If it is dinnertime, press 1."

"If your idea of dinnertime is 3:13 P.M., press 2," the machine said. I did not press 2. Nobody eats dinner at 3:13 in the afternoon. The machine said, "If your idea of dinnertime is one of those many points in time between 4 and 5 P.M., press 3."

When I did not press 3 the machine said, "If your idea of dinnertime is encompassed within the time frame created by 5 P.M. and 6 P.M., press 4."

Impatient, I pressed 4. The machine said: "That is an inappropriate response. If you wish to know why, press 1."

I pressed 1. The machine said, "The period between 5 P.M. and 6 P.M. is not dinnertime. It is suppertime. Press 9."

Give us this day our daily bread.

I pressed 9 and was immediately disconnected again. I decided to write my message to Taffy and send it through the mail. It looked like the fast way to get through now that the telephone had been so perfected that it was practically useless.

Of course it had been so long since I had written a letter that I could barely remember how to do it.

"Dear Taffy," I wrote, "If you want to learn something interesting, press this letter once and return it to me in the enclosed self-addressed, stamped envelope. If you do not want to learn something interesting, do not press this letter but return it to me anyhow."

Five days later my letter was returned, but I could not tell whether Taffy had pressed it once or not at all. Paper is funny that way. Pressing doesn't leave a lasting impression. Clay was what I needed. Soft modeling clay.

"Dear Taffy," the clay said, and repeated the instructions sent earlier on paper. Off it went parcel post.

Three days later my phone rang. "This is Taffy calling," said a machine. "If you would like to hear what Taffy has to say, press 1."

I pressed 1. Taffy's machine said: "Since Taffy believes it is rude to speak to human beings through a machine, I am unable to let you know what Taffy has to tell you. I can, however, speak this message to another machine. If you would like to put your machine on the phone, press 3."

This so enraged me that I slammed down the phone without pressing anything at all. The phone promptly rang again. It was Taffy's machine calling back. "That was not a valid response," the machine said. "If you would like to make a valid response, press 4."

I pressed 4. "That is much better," said Taffy's machine. "In this age of instant communications, rudeness cannot be tolerated on the telephone. If you would like me to give you another chance instead of faxing a note home to your mother, press 5."

I pressed 5. "If you want to thank me, press 6," said Taffy's machine.

I pressed 6, and the machine said, "Now if you would like to know how to hear what Taffy has to say, press 7."

I pressed 7. The machine said, "In your manifestation as a human being, telephone Taffy this evening at dinnertime."

I waited until precisely 8 P.M., having been persuaded ever since seeing *Dinner at Eight* that this was the one and only dinnertime for persons vital enough to own industrial empires and telephone-answering machines.

Naturally, a machine answered. "If it is dinnertime, press 1," the machine said. I hesitated. I had forgotten what I'd wanted to tell Taffy. Fury had its way with me. I pressed the entire phone underfoot a dozen times. "That is not a valid response," said the fragments, departing through the window.

—*Russell Baker,* New York Times, *June 26, 1993*

Say hello to the Hobbit microprocessor, the universal phone card, stored image, fiber-optic transmission, E-mail, PhoneLink, TrueVoice, EasyReach, Receptor, Embarc, teleconferencing, and code division multiple access. The interactive future is here, with the sky the limit and mega-mergers happening right and left between cable and telephone companies.

"You come home from work and grab the remote," *Newsweek* reported in 1993, looking at how life almost is. "As you putter around, removing tie or pantyhose, and occasionally checking the picture, your personal video navigator brings you up to date. You find out what TV shows the kids watched after school, and hear a reminder from the florist: It's time to send Aunt Agnes's birthday bouquet; how about this arrangement? You look at a copy of Tommy's report card, issued that day, and a list of movies you could watch that night, based on how much you loved *The Age of Innocence.* You click on the beef bourguignon how-to that you selected this morning; you've got all the ingredients because the program automatically faxed a list to Safeway, which delivered."

"ALTHOUGH THERE IS A GREAT FIELD FOR THE TELEPHONE in the immediate present, I believe there is still greater in the future," Alexander Graham Bell wrote in 1878. Little did he know what wonders his wonder would wrought. Little could he imagine that making a plain, old-fashioned call would become the maddening chore of choosing from an electronic menu. His dream of a telephone for Everyman has come true. But in its realization, the basic connection has been

---

I understand why the telephone is in theory a good thing, and I would be very unhappy were I unable to make use of one when I pleased, or in an emergency, but the truth is that I do not like the telephone. I do not like the sound of it. I do not particularly admire how they look, even the old ones, even the new ones that look like Barbie or Mickey Mouse. I do not like what seems to be happening, in a sociological sense, with telephones—that people on the street or in a restaurant now take out cellular phones to make a quick call during dinner. I do not think that answering machines have made my life any simpler or better. They may have improved the lives of some people, but not mine. It now takes six calls to impart the kind of information that used to be accomplished in one telephone conversation; there are three or four exchanges of messages, then the actual information is left on the machine, and then there is a later message in which the caller seeks to know if his message has been received, and if it is agreeable. I understand that the supposed benefit of answering machines is that you do not have to have a conversation with another human being, that that is the very point of them, not efficiency, but that does not seem to me to be a good thing either. I would prefer neither to talk on the telephone nor to leave a message or to receive one.

There is one time, however, that I do like the telephone, and that is at the beginning of a love affair when the telephone becomes a totem. It is possessed of magic, symbolism, power. The telephone becomes the source of all pleasure and pain. One stares at it. Lifts the receiver to make sure that it is working. Restrains oneself from making other calls. Holds it in one's lap. Throws it across the room.

—Susanna Moore, novelist, 1993

The line of a telephone subscriber is like a hallway leading from the house, or office, where he has his telephone instrument, to the central office of his telephone city. And there he has a door opening onto all the national highways of speech. Through this telephone door and along the hallway, formed by the wires which connect his instrument to the central office, conversations can take place with anyone else who also has a door on the national highway.

—*John Mills,* The Magic of Communication, *1925*

Dear Díary:

Saturday afternoon, I am wandering around Bloomingdale's. In the Donna Karan section I spot a shopper considering blazers. In addition to the one she is trying on, others are draped around her as if she were a human hanger.

With the gracefulness of a runway model she twirls around the mirrors; with the vocal chords of a Loehmann's shopper she beckons the saleswomen: "Which one matches my pants? Do these go together?"

Uncertain of the communal response, she reaches into her purse and brings forth a portable phone. She dials a number.

Frantically, she speaks: "Hi, honey. I'm in Bloomie's and can't make a decision. Should I do a colorful linen, a solid or a beige?"

Silence. And then, angrily: "No, no. I can't go to Saks. You know I can't call you from there. Don't you remember? I get horrible phone reception in their sportswear department."

I head for Saks.

—*Meredith Wollins, "Metropolitan Diary,"* New York Times, *March 24, 1993*

broken. The caller is often at the whim of a robot, spewing out heartless sounds and relentless instructions. The long-distance "operator" with a recorded spiel allows no collect call to go through until an exact reply is given in an allotted minisecond. The answering machine screens unwelcome callers, and the telephone company's own Call Answering service eliminates the very answering machine itself.

The instrument that once brought people together has spawned an industry that now keeps them apart, and the gossip's favorite toy has become a tool of desocialization. Somehow, the original premise of person-to-person has been sabotaged. And surely none of this is anything at all like what Alexander Graham Bell had in mind when he held his first tuning fork and dreamed of the human voice magically transmitted through the air over a wire.

(Right) From here to eternity: tombstone in a southern cemetery.

Permissions and Acknowledgments

The following pages of permissions and acknowledgments constitute a continuation of the copyright page found at the front of the book. While every effort has been made to trace all present copyright holders of the material in this book—whether companies or individuals—any unintentional omission is hereby apologized for in advance, and we should of course be pleased to correct any errors in acknowledgments in any future edition of this book.

p.2 (photo): Library of Congress; **p.6** (illustration, bottom): Library of Congress; **p.7** (photo): Courtesy of AT&T Archives; **p.9** (illustration): Library of Congress; **p.13** (photo): Library of Congress; **p.14** (advertisement at left): Courtesy AT&T Archives; **p.16:** Courtesy of Hallmark Cards; **p.19** (photo, top left): Courtesy of AT&T Archives; **p.23** (text, right): Reprinted by permission of Mary Finch Hoyt; **p.25** (text, right): From *Working* by Studs Terkel. Copyright © 1972, 1974 by Studs Terkel. Reprinted by permission of Pantheon Books, a division of Random House, Inc.; **p.27** (text): All Rights Reserved. Copyright © 1957 by Jerome Lawrence and Robert E. Lee. Copyright renewed 1985 by Jerome Lawrence and Robert E. Lee; **p.29:** "Manual System" from *Early Moon* by Carl Sandburg and illustrated by James Daugherty. Copyright 1920 by Harcourt Brace & Company and renewed 1948 by Carl Sandburg. Reprinted by permission of the publisher; **p.35** (text, bottom): Copyright © 1960 by Holt Rinehart & Winston. Reprinted by permission of Ellen N. Baldinger; **p.37** (photo): Courtesy of AT&T Archives; **p.38:** "Under a Telephone Pole" from *Chicago Poems* by Carl Sandburg. Reprinted by permission of Harcourt Brace & Company; **p.41** (photos, right): Reprinted by permission of Henry Dreyfuss Associates; **p.47** (text): Reprinted by permission of Jonathan Schwartz; **p.52** (text): Reprinted with permission of Gertrude Goodrich Snyder; **p.53** (text, top left): From *Goodbye, Columbus,* by Philip Roth. Copyright © 1959, renewed 1987 by Philip Roth. Reprinted by permission of Houghton Mifflin Company. All rights reserved; **p.54** (text): Reprinted by permission. © 1946, 1974 The New Yorker Magazine, Inc. All rights reserved; **p.55** (text, top right): From *Enter Talking* by Joan Rivers. Copyright © 1986 by Joan Rivers. Used by permission of Dell Books, a division of Bantam Doubleday Dell Publishing Group, Inc.; **p.60** (text, left): From *Life with Father* by Clarence Day. Copyright © 1948, Alfred A. Knopf, Inc.; **p.70** (text, top center): From *How to Make Yourself Miserable,* by Dan Greenburg and Marcia Jacobs. Copyright © 1966, Random House, Inc.; **p.71** (text, left): From *Among the Porcupines,* by Carol Matthau. Copyright © 1992, Turtle Bay Books; **p.75** (text): From *Vox,* by Nicholson Baker. Copyright © 1992, Alfred A. Knopf, Inc.; **p.76** (text): From "Fear of Phoning," by Bruce Feirstein originally appeared in *Mademoiselle;* **p.78** (photo): Photofest; **p.78** (text, right): Reprinted by permission of Jan Hodenfield; **p.80** (text): From *Mayflower Madam* by Sydney Biddle Barrows. Copyright © 1986 by Sydney Biddle Barrows and William Novak. By Permission of William Morrow, Inc.; **p.82** (text): From *Johnny Got His Gun,* by Dalton Trumbo. Copyright © 1939, 1951, 1991 by Dalton Trumbo. Published by arrangement with Carol Publishing Group. A Citadel Underground Book; **p.84** (text): "A Telephone Call," from *The Portable Dorothy Parker.* Introduction by Brendan Gill. Copyright 1928, renewed © 1956 by Dorothy Parker. Used by permission of Viking Penguin, a division of Penguin Books USA Inc.; **p.85** (text, top): From *How I Got To Be Perfect* by Jean Kerr. Copyright © 1978 by Collins Productions. Inc. Used by permission of Doubleday, a division of Bantam Doubleday Dell Publishing Group, Inc.; **p.85** (photo, top left): Photofest; **p.88** (illustration): From the *Wall Street Journal*—Permission, Cartoon Features Syndicate; **p.88** (text): Copyright © 1975 by Michael Korda and Paul Gitlin. Trustee for the benefit of Christopher Korda. Reprinted by permission of Michael Korda; **p.90** (photo, top right): Courtesy of the John F. Kennedy Library; **p.91** (text): Copyright © 1974 by Carl Bernstein and Bob Woodward. Reprinted by permission of Simon & Schuster, Inc.; **p.92** (text, bottom): Excerpted with permission from "Dr. Jekyll-Hyde Is on the Phone" by Don Herold, *Reader's Digest,* August 1937. Copyright © 1937 by the Reader's Digest Assn., Inc. Reprinted by permission of Doris Lund; **p.93** (text): From *She's Off to Work: A Guide to Successful Earning and Living,* by Gulielma Fell Alsop and Mary Francis McBride. Copyright © 1941 by The Vanguard Press, Inc., Copyright © renewed 1969 by Gulielma Fell Alsop and Roberta McBride. Reprinted by permission of Vanguard Press, a division of Random House, Inc.; **p.97** (text, center left): From *Mr. Personality,* by Mark Singer. Copyright © 1988, Alfred A. Knopf, Inc.; (text, bottom): From *The Great Goldwyn,* by Alva Johnston. Copyright © 1937, Random House, Inc.; (text, top): Excerpt from Chapter 14 of *The Studio* by John Gregory Dunne. Copyright © 1969 by John Gregory Dunne. Reprinted by permission of Farrar, Straus & Giroux, Inc.; **p.98** (illustration): Illustration by James McMullan; (text, top): *The Front Page* by Ben Hecht and Charles MacArthur. Copyright © 1928 by Cayben Productions and Charles MacArthur. Copyright © 1928 by Ben Hecht and Charles MacArthur. Copyright © 1950 (acting edition) by Ben Hecht and Charles MacArthur. Copyright © 1955 (in renewal) by Ben Hecht and Charles MacArthur; **p.99** (text): From *A Rock in Every Snowball,* by Frank Sullivan. Reprinted by permission of the Historical Society of Saratoga Springs, Saratoga Springs, New York; **p.101** (poem, bottom): From *Now and Then: Poems 1976–1978* by Robert Penn Warren. Copyright © 1976, 1977, 1978 by Robert PennWarren. Reprinted by permission of Random House, Inc.; (text, top left): From *Zolar's Encyclopedia and Dictionary of Dreams,* by Zolar. Copyright © 1963 by Zolar Publishing Company, Inc. Used by permission of Doubleday, a division of Bantam Doubleday Dell Publishing Group, Inc.; **p.102:** © The Curtis Publishing Company; **p.105** (text): From *How to Behave—Though a Debutante* by Emily Post. Illustrated by John Held. Copyright 1928 by Doubleday. Used by permission of Doubleday, a division of Bantam Doubleday Dell Publishing Group, Inc.; **p.107** (illustrations, top right): Courtesy of Hallmark Cards; **p.110** (photo): Photofest; (sheet music): From *The Telephone Hour,* by Charles Strouse and Lee Adams. Copyright © 1960 (Renewed 1988) Lee Adams and Charles Strouse c/o The Songwriter's Guild. Used by Permission of CPP/BELWIN, INC., Miami, FL. All Rights Reserved; **p.111** From *The Temple of Gold,* by William Goldman. Copyright © 1957, 1985 by

William Goldman. Used by permission of the author. All rights reserved; **p.112** (text): From *Youth: The Years from Ten to Sixteen,* by Arnold Gesell, M.D., Frances L. Ilg, M.D., and Louise Bates Ames, Ph.D. Copyright © 1956 by Harper & Row. Reprinted by permission of Louise Bates Ames, Ph.D., The Gesell Institute of Human Development; **p.113** (text, top): From *When Your Child Drives You Crazy,* by Eda LeShan. St. Martin's Press, Inc., New York, NY. Copyright © 1985 by Eda LeShan; **p.114** (text): Reprinted by permission of Carol Rinzler. Copyright 1981; **p.115** (illustration, bottom right): Courtesy of Hallmark Cards; **p.116:** © The Curtis Publishing Company; **p.119** (text): From *Last Respects,* by Jerome Weidman. Copyright 1971 by Jerome Weidman. Copyright renewed © 1983 by Jerome Weidman. Reprinted by permission of Brandt & Brandt Literary Agents, Inc.; **p.120** (illustration, top left): Courtesy of AT&T Archives; **p.121** (photos, top left and bottom left): Photofest; **p.122** (text): From *Strangers When We Meet,* by Evan Hunter. Copyright © 1958 by Evan Hunter. Copyright © renewed 1986 by Evan Hunter; **p.123** (text, bottom): From "The Master of the Universe" from *The Bonfire of the Vanities* by Tom Wolfe. Copyright © 1987 by Tom Wolfe. Reprinted by permission of Farrar, Straus & Giroux, Inc.; **p.124** (text, bottom): Reprinted by permission of Curtis Brown, Ltd. Copyright © 1942 by A.J. Liebling; **p.124:** "To a Lady in a Phone Booth," copyright © 1958 by Phyllis McGinley, from *Times Three,* by Phyllis McGinley. Used by permission of Viking Penguin, a division of Penguin Books USA Inc.; **p.125** (photo): The Bettmann Archive/Bettmann Newsphotos; **p.126** (illustration): Copyright © 1994 The Estate of Keith Haring; **p.127:** Drawing by Bernard Schoenbaum. Copyright © 1993 The New Yorker Magazine, Inc.; **p.128** (text, top): Reprinted by permission of Ralph Schoenstein; **p.129** (text, right): Copyright © 1988, Time, Inc. Reprinted by permission; **p.131** (text): Copyright © 1993 by the New York Times Company. Reprinted by permission; **p.132:** Photo by Sara Barrett; **p.133** (text, right): Copyright © 1993 by the New York Times Company. Reprinted by permission; (photo, right): Scoring in Heaven Collection, Lucinda Bunnen and Virginia Warren Smith, photographers. Courtesy of the Center for the Study of Southern Culture.

Bibliography

American Telephone and Telegraph Company, *Alexander Graham Bell.* New York, 1947.

Aronson, Sidney H. "Bell's Electrical Toy." *The Social Impact of the Telephone.* Edited by Ithiel de Sola Pool. Cambridge, M.A.: MIT Press, 1977.

von Auw, Alvin. *Heritage and Destiny: Reflections on the Bell System in Transition.* New York: Praeger, 1983.

Baida, Peter. "Breaking the Connection." *American Heritage.* June/July, 1985.

Barnett, Lincoln. "The Voice Heard Round the World." *American Heritage.* April, 1965.

Bates, Louise and Frances L. Ilg. *Your Four Year Old.* New York: Delacorte, 1976.

Boettinger, H.M. *The Telephone Book.* Croton-on-Hudson, N.Y.: Riverwood Publishers Ltd., 1977.

Brooks, John. *Telephone: The First Hundred Years.* N.Y.: Harper & Row, 1975.

Brown, Henry Collins, editor. *Valentine's Manual of Old New York.* No. 5, New Series. New York: Valentine's Manual Inc., 1921.

Bruce, Robert V. *Bell: Alexander Graham Bell and the Conquest of Solitude.* Boston: Little, Brown 1973.

Casson, Herbert N. *The History of the Telephone.* Chicago: A.C. McClurg & Co., 1910.

Chandler, Raymond. *The Long Goodbye.* Boston: Houghton Mifflin, 1954.

Cousins, Margaret. *The Story of Thomas Alva Edison.* New York: Landmark Books, Random House, 1965.

de Camp, L. Sprague. *The Heroic Age of American Invention.* Garden City: Doubleday, 1961.

de Sola Pool, Ithiel, editor. *The Social Impact of the Telephone.* Cambridge, M.A.: The MIT Press, 1977.

Dooner, Kate E. *Telephone Collecting: Seven Decades of Design.* Atglen, P.A.: Schiffer Publishing Ltd., 1993.

Eco, Umberto and G.B. Zorzoli. *The Picture History of Invention.* New York: Macmillan, 1963.

Fagen, M.D., editor, et al. *A History of Engineering and Science in the Bell System.* Bell Telephone Laboratories, Inc., 1975.

Flatow, Ira. *They All Laughed. From Light Bulbs to Lasers: The Fascinating Stories Behind the Great Inventions That Have Changed Our Lives.* New York: HarperCollins, 1992.

Grigson, Geoffrey. *Things.* New York: Hawthorn Books, Inc., 1957.

Hackenburg, Herbert J., Jr. *Muttering Machines to Laser Beams.* Denver: Mountain Bell, 1986.

Hapgood, Fred. "At 411, It's Simply a Matter of Keeping in Tune with the Numbers." *Smithsonian,* November, 1986.

Heyn, Ernest V. *Fire of Genius: Inventors of the Past Century.* Garden City: Anchor Press/Doubleday, 1976.

Hine, Thomas. *Populuxe.* New York: Knopf, 1986.

Hylander, C.J. *American Inventors.* New York: Macmillan, 1934.

Kutner, Lawrence. *Parent & Child.* New York: Morrow, 1991.

Maddox, Brenda. "Women and the Switchboard," *The Social Impact of the Telephone.* Edited by Ithiel de Sola Pool. Cambridge, M.A.: The MIT Press, 1977.

Patton, Phil. *Made in U.S.A.: The Secret Histories of the Things That Made America.* New York: Grove Weidenfeld, 1992.

Sheth, Jagdish N. and David A. Heffner. *Voice with a Smile: True Stories from American Telephone Operators.* Barrington, I.L.: PERQ Publications, 1991.

Todd, Kenneth P., Jr., editor. *A Capsule History of the Bell System.* American Telephone and Telegraph Company [n.d.].

Updike, John. "Telephone Poles." From *Telephone Poles and Other Poems.* New York: Knopf, 1963.

Waite, Helen Elmira. *Make a Joyful Sound.* Philadelphia: Macrae Smith Company, 1961.

Watson, Thomas A. "The Birth and Babyhood of the Telephone." American Telephone and Telegraph Company, 1913.